AI Based Solutions for Inclusive Quality Education

AI Based Solutions for Inclusive Quality Education helps readers discover practical AI tools for diverse learning styles and abilities, fostering inclusivity in education and maximizing the individual potential of students and educators. It provides cutting-edge solutions designed to remove barriers due to accessibility and create an equitable classroom. Through practical applications and case studies, the book explores the power of AI in creating personalized learning pathways and methods of engagement in the classroom while emphasizing that the human touch, with empathy and critical thinking, remains central to the learning process. It addresses concerns about the ethical development and deployment of AI in education.

Features

- Discusses a topic that is at the forefront of concerns in the education sector with the disruptive impact of advances in AI.
- Provides explanations for specific AI platforms and tools, along with their features and applications in education.
- Includes case studies and discusses inclusivity and the lessons learned from each AI application.
- Explains how to ethically handle bias, privacy and transparency in AI-based solutions in education.
- Serves as a comprehensive guide for educators, policymakers and technology developers interested in the power of AI in educational environments.

This book serves as a timely reference in the current era of AI tools and education. It is written for those who work in the education sector, as well as researchers, academicians, undergraduate and graduate students, those involved with undergraduate teaching in critical thinking and problem-solving for information systems, information technology, computer science and engineering and all teaching institutions where AI tools have been or will be introduced.

Higher Education and Sustainability

J. Paulo Davim, *Professor, Department of Mechanical Engineering, University of Aveiro, Portugal*

This new series fosters information exchange and discussion on higher education for sustainability and related aspects, namely academic staff and student initiatives, campus design for sustainability, curriculum development for sustainability, global green standards: ISO 14000, green computing, green engineering education, index of sustainability, recycling and energy efficiency, strategic sustainable development, sustainability policies, sustainability reports, etc. The series will also provide information on principles, strategies, models, techniques, methodologies, and applications of higher education for sustainability. It aims to communicate the latest developments and thinking as well as the latest research activity relating to higher education, namely engineering education.

Higher Education and Sustainability
Opportunities and Challenges for Achieving Sustainable Development Goals
Edited by Ulisses Manuel de Miranda Azeiteiro and J. Paulo Davim

Designing an Innovative Pedagogy for Sustainable Development in Higher Education
Edited by Vasiliki Brinia and J. Paulo Davim

Higher Education
Progress for Management and Engineering
Edited by Carolina Machado and J. Paulo Davim

Principles of Entrepreneurship in the Industry 4.0 Era
Edited by Rajender Kumar, Rahul Sindhwani, Tavishi Tewary, and J. Paulo Davim

Technopreneurship and Sustainability
Innovation, Challenges, and Opportunities
Edited by Rajender Kumar, Rahul Sindwani, Raman Kumar, Punj Lata Singh, and J. Paulo Davim

AI Based Solutions for Inclusive Quality Education
Edited by K.M. Soni, Nitasha Hasteer, Aditi Bhardwaj, Rahul Sindhwani and J. Paulo Davim

For more information about this series, please visit: www.crcpress.com/Higher-Education-and-Sustainability/book-series/CRCHIGEDUSUS

AI Based Solutions for Inclusive Quality Education

Edited by
K.M. Soni, Nitasha Hasteer, Aditi Bhardwaj,
Rahul Sindhwani and J. Paulo Davim

CRC Press
Taylor & Francis Group
Boca Raton London New York

CRC Press is an imprint of the
Taylor & Francis Group, an **informa** business

Designed cover image: Shutterstock

First edition published 2026
by CRC Press
2385 NW Executive Center Drive, Suite 320, Boca Raton FL 33431

and by CRC Press
4 Park Square, Milton Park, Abingdon, Oxon, OX14 4RN

CRC Press is an imprint of Taylor & Francis Group, LLC

ISBN: 9781032909783 (hbk)
ISBN: 9781032909790 (pbk)
ISBN: 9781003560814 (ebk)

DOI: 10.1201/9781003560814

Typeset in Palatino
by Newgen Knowledge UK

Contents

Preface

Education is the very root of human intelligence, it plays the most crucial role in developing and shaping human intellect by feeding it with knowledge, helping it build problem-solving skills and reasoning capabilities. Education is essential for an individual for personal as well professional growth, leading them to transform society with remarkable innovations.

In the modern age, the fast-paced technologies like artificial intelligence (AI) are paving the way for education's new era. Education has not been this accessible, personalized and flexible before. This transformational shift has resulted in inclusive education for each and every learner across the globe irrespective of their background, making it more than an advancement of existing technologies. This book explores how AI can be harnessed to create inclusive quality education for all and is a compilation of chapters from leading academicians revolving around the agenda of AI and its transformative role in the education sector. This book starts with an Introduction to the theme – 'Transformative Power of Artificial Intelligence in the Educational Context'. Following the introductory chapter, the next chapter 'Leveraging AI to Foster Inclusive Quality Education in Higher Educational Institutes' discusses how AI-powered learning benefits educators and learners along with the identification of modern AI tools used in higher education. 'AI and Inclusive Education: Enhancing Equity, Engagement, and Excellence', further strengthens the compilation by highlighting how AI and modern technologies have opened gates for all learners irrespective of their background or skill sets. Following this, the next chapter, 'Artificial Intelligence Tools for Instructors and Learners to Optimize the Teaching and Learning Processes' delves into powerful solutions to optimize the learning and teaching process by leveraging AI-based intelligent tutoring and learning assistants and lists the various AI-enabled platforms. 'Artificial Intelligence – A Digital Architect for Curriculum Development' highlights the use of AI for increased efficiency in curriculum design, AI for personalized learning and the function of generative AI in curriculum development and ethical considerations in AI driven curriculum design. The chapter on 'Artificial-Intelligence-Based Automated Instructor Feedback System for Education and Research' presents a systematic review of literature on automatic feedback and AI-based automated feedback systems for education to enhance the quality of teaching. The next chapter on 'Enhancing Teacher Preparation for Integrating Computational Thinking through Unplugged Activities: Utilizing Learning Styles and MBTI Personality Types', discusses a potential solution for the training of educators based on their learning styles to achieve effective teaching performance by addressing the distinct needs and challenges faced in classrooms. Finally, 'Revolutionizing AI to Break Down Barriers in Education for Children with Special Needs' explores innovative AI-related prospects for creating a barrier0free educational environment for children with special needs.

We, as editors, take this opportunity to thank our organizations, which have provided us with a conducive environment to publish this book. We wish to acknowledge all the contributors of different chapters who cooperated with us at every stage of publication and helped us to sail through this mammoth task. We would also like to extend our gratitude to the expert reviewers who have helped us in reviewing the book chapters. We owe our sincere gratitude to all our family members and friends who have helped us through this journey of publishing a book. Our appreciation goes to each and every individual who has supported us in this endeavour, with a special mention to Mr. Lalit Sharma. Last but not least, we are grateful to the publishing team of CRC Press who have provided all the support to us in the compilation of the book and shaped it into a marketable product.

We hope that this book will prove to be a valuable resource for AI enthusiasts aspiring to gain knowledge in the field of education technologies and those who envision a future where AI serves as a tool for inclusivity and not exclusion.

About the Editors

K.M. Soni gained his B.E. in Electrical Engineering & M.E. in Control and Instrumentation from Motilal Nehru National Institute of Technology Allahabad and PhD in Electrical Engineering from Jamia Millia Islamia, A Central University, New Delhi. Prof. Soni is a Professor at Amity School of Engineering & Technology, Dy. Dean (Engg. & Tech.), in-charge of PhD Dept. AUUP. He has completed many online courses related to Education Technologies such as NBA Accreditation and Teaching and Learning in Engineering (NATE), Teaching and Learning in Engineering (TALE), Designing Learner-Centric MOOCs, Outcome-Based Pedagogic Principles for Effective Teaching, etc. He has over 25 years of teaching, research, and academic administrative experience. He has authored books on Circuits and Systems, Signals and Systems, Basic System Analysis, Network Analysis and Synthesis, and Advanced Control Systems. He is the author/co-author of over 50 technical research papers in refereed conferences and journals and has served as a reviewer, programme committee member, and session chair of national and international conferences in India and abroad. He is an Assessor of the National Assessment and Accreditation Council (NAAC) and visits government and private universities/institutions as a peer-team member. He is a Life Member of the Indian Society for Technical Education (ISTE), Institution for Electronics and Telecommunication Engineers (IETE), Senior Member of IEEE (USA), Member of CSI, IET (U.K.), IUCEE, IoT Society of India, and other similar organizations. His current research interests are in Power Electronics, Power Systems, Advanced Control Systems, Signal Processing, IoT, Artificial Intelligence, Education Technologies, and other related areas.

Nitasha Hasteer is a Professor and Head of the Information Technology Department, as well as the Deputy Director (Academics) at the Amity School of Engineering & Technology, Amity University, Uttar Pradesh. She holds a PhD in Computer Science and Engineering and has over 24 years of teaching, research, and administrative experience across academia and industry. Her primary research interests lie in Machine Learning, Cloud Computing, Crowdsourced Software Development, Software Project Management, and Process Modelling using Multi-Criteria Decision-Making Techniques. Prof. Hasteer has made significant contributions to these fields, publishing more than 80 research papers in renowned international journals and conferences. Her work focuses on advancing computational techniques and methodologies that improve software development processes and decision-making frameworks, as well as exploring the integration of machine learning with cloud-based systems for optimized, scalable solutions. She has successfully guided numerous postgraduate dissertations and is currently mentoring PhD scholars in areas related to the Computer Science and Engineering and Management domains. Prof. Hasteer has also contributed to the academic community as an editor for books published by prestigious publishers. Her academic leadership extends to the broader research community, where she has served on the editorial boards of several high-impact international conferences and journals. She has secured research funding from prominent governmental organizations, including the Science and Engineering Research Board (SERB), the Department of Science & Technology (DST), the Defence Research & Development Organization (DRDO), the Indian National Science Academy (INSA), and the Council of Scientific & Industrial Research (CSIR) for organizing international conferences that advance knowledge in the domain of Information Technology.

Aditi Bhardwaj obtained her BE in Electrical & Electronics Engineering from Rajiv Gandhi Proudyogiki Vishwavidyalaya Bhopal, Madhya Pradesh, and MTech in Modeling and Simulation from Savitribai Phule Pune University, Pune, Maharashtra, and PhD in Computer Science and Engineering from Gautam Buddha University, Greater Noida, Uttar Pradesh. Prof. Aditi is an Assistant Professor in the Computer Science and Engineering Department at Amity University, Noida, Uttar Pradesh. With over five years of teaching and research experience, Prof. Aditi Bhardwaj has established herself as a respected authority in Artificial Intelligence and Machine Learning. Her research contributions have been widely published in prestigious publishers including IEEE, Elsevier, Taylor & Francis, etc., solidifying her expertise. Additionally, she has a collaborative spirit and works as an Editor and Co-editor of various books with publishers such as Taylor & Francis, IGI Global, etc. Her core competencies lie in Artificial Intelligence, Machine Learning, Emotion Recognition, and EEG Signals. Prof. Aditi Bhardwaj possesses a deep understanding of these complex fields and actively contributes to developing innovative frameworks and concepts, furthering the advancement of knowledge.

Rahul Sindhwani is currently serving as an Assistant Professor at the Indian Institute of Management (IIM), Sambalpur, Odisha, India. Prior to this, he was associated with the Birla Institute of Management and Technology, Greater Noida. Prof. Sindhwani holds a Bachelor of Technology (BTech) degree in Mechanical Engineering and was awarded a gold medal during his Master of Technology (MTech) in Industrial Engineering and Management. He earned his PhD in Industrial Engineering and Management from J.C. Bose University of Science and Technology and subsequently completed his Post Doc in Operations Management and Decision Science area from the Indian Institute of Management (IIM), Amritsar. His areas of expertise include the Digitalization of Supply Chains, Agile Project Management, and

Operations Management. With more than a decade of experience in teaching and research, Dr. Sindhwani has published extensively in internationally recognized ABDC- and ABS-ranked journals, such as *Annals of Operations Research, IEEE Transactions on Engineering Management, Journal of Cleaner Production*, and *Technological Forecasting and Social Change*, among others. In addition to his research contributions, Dr. Sindhwani serves as a guest editor for the *Journal of Global Responsibility*. He is also the editor of five books on the industrial engineering and management area, published by Springer and Taylor & Francis. He has developed innovative frameworks and concepts that have significantly improved organizational operational efficiencies and competitiveness, showcasing his deep expertise in these fields.

J. Paulo Davim is Professor at the University of Aveiro, Portugal. He is also distinguished as honorary professor in several universities/colleges/institutes in China, India and Spain. He received his PhD in Mechanical Engineering, MSc in Mechanical Engineering (materials and manufacturing processes) and Mechanical Engineering degree (5 years), from the University of Porto (FEUP), the Aggregate title (Full Habilitation) from the University of Coimbra and the DSc (Higher Doctorate) from London Metropolitan University. He is Fellow (FIET) of IET-London, Fellow (FIMF) of IMF-Birmingham and Eur Ing by Engineers Europe FEANI-Brussels. He is also Senior Specialist Engineer by the Portuguese Order of Engineers with an MBA and Specialist titles in Industrial Management as well as in Metrology. He has more than 35 years of teaching and research experience in Manufacturing, Materials, Mechanical and Industrial Engineering, with special emphasis in Machining, Tribology & Surface Engineering. He has also interest in Design, Management, Sustainability, Industry 5.0, Engineering Education and Higher Education for Sustainability. He has guided large numbers of postdoc, PhD and master's students as well as has coordinated and participated in several financed research projects. He has received several scientific awards and honors. He has worked as evaluator of projects for ERC-European Research Council and other international research agencies as well as examiner of PhD thesis for many universities in different countries. He is the Editor in Chief of several international journals, book Series Editor and Scientific Advisory for many conferences.

Contributors

Harshit Bhardwaj
Amity University
Uttar Pradesh, Noida, India

Taranjeet Duggal
Manav Rachna International Institute of Research and Studies
Faridabad, Haryana, India

Ankit Dubey
SCSET
Bennett University
Greater Noida, India

Umesh Gupta
SCSET
Bennett University
Greater Noida, India

Nitasha Hasteer
Amity University Uttar Pradesh
Noida, India

Srishti Singh Kaira
Manav Rachna International Institute of Research and Studies
Faridabad, Haryana, India

Sangeeta Khorana
Aston University
Birmingham, England, United Kingdom

Neha Kumari
O.P. Jindal Global University
Haryana, India

Rajesh Kumar Modi
Stuvalley Technologies Pvt. Ltd, India

Harshita Mogha
Amity University Uttar Pradesh, Noida, India

Bhairab Chandra Patra
Indian Institute of Management – Sambalpur
Orissa, India

Ayushman Pranav
SCSET
Bennett University
Greater Noida, India

Janakarajan Ramkumar
Indian Institute of Technology
Kanpur, Uttar Pradesh, India

Rahul Sagwal
Indian Institute of Technology
Kanpur, Uttar Pradesh, India

Abhinav Sharma
Indian Institute of Technology
Kanpur, Uttar Pradesh, India

Rahul Sindhwani
Indian Institute of Management – Sambalpur
Orissa, India

K. M. Soni
Amity University Uttar Pradesh
Noida, India

Pradeep Tomar
Department of Computer Science and Engineering
University School of Information and Communication Technology
Gautam Buddha University
Greater Noida, U.P., India

Shreya Tyagi
Manav Rachna International Institute of Research and Studies
Faridabad, Haryana, India

Pallavi Upadhyay
Department of Computer Science and Engineering
University School of Information and Communication Technology
Gautam Buddha University
Greater Noida, U.P., India

Vinita Yadav
Department of Computer Science and Engineering
University School of Information and Communication Technology
Gautam Buddha University
Greater Noida, U.P., India

1

Transformative Power of Artificial Intelligence in the Educational Context

K.M. Soni, Nitasha Hasteer, Rahul Sindhwani, Harshita Mogha and Sangeeta Khorana

1.1 The Significance of Education

Education has long been regarded as the very foundation of human civilization and as the primary tool for empowerment and social development. Today, it is more relevant than ever, in view of the unprecedented global challenges and opportunities in this fast-changing world. The power of education cuts across multiple axes of human development. Quality education is the great counterbalance in enhancing social mobility and economic progression across the different divisions of the socioeconomic ladder [1]. Access to quality education can lead to better job opportunities with higher income and significantly improving social mobility metrics. There is a strong correlation between the level of education and better health, civic engagement, and living standards.

While education benefits individuals on a large scale, it has a more profound impact on the development of society [2]. Countries with greater educational attainment have consistently shown stronger economic growth, improved public health metrics, and greater technological innovations. Yet, the current global educational landscape poses major challenges to achieving inclusive quality education. An estimated 258 million children and youth are still out of school, with inequalities in access to education being widest in developing regions [3]. This results in a lack of opportunities and advancements for millions of young learners. The many barriers to access include socioeconomic inequalities, gender disparities, and geographic barriers. Further amplifying these challenges are emerging global issues. Changes in climate, political instability, and technological disruption may further impede access to and quality of education for underprivileged groups, as the inequality cycle in education continues to worsen. These are deep-rooted issues that require comprehensive and innovative solutions. To achieve innovative and transformational changes in the world, the most powerful tool is still,

DOI: 10.1201/9781003560814-1

unquestionably, education. The old approaches to education are insufficient to meet these needs. It is not just new technologies that educational systems need, but also new pedagogies to close the educational gaps that guarantee equal-access opportunities to quality learning experiences.

1.2 The Role of Technology in Education

Technology, with regard to education, is arguably one of the most exciting and transformative forces currently redefining modern pedagogy. For the last few decades, this has played out through a sequence of innovation and adaptation phases marking significant transitions in educational paradigms [4,5]. It shows changes more profound than those of mere technology, reflecting an understanding of the basics of how people learn and teach.

The historical trajectory of educational technology is a very interesting narrative of progressive innovation [5]. From the introduction of audiovisual aids in the mid-20th century to modern digital tools, every technological advancement has changed the educational landscape fundamentally. Film strips and overhead projectors made up the early technological intervention that laid the foundation for more sophisticated digital tools like smart teaching. Technology serves a pedagogical need and also continues to push the boundaries of what is possible in education [6].

Digital learning environments are the results of many decades of technological development in academia. Today's learning management systems have evolved from rather simple content-delivery systems that have now become highly elaborate ecosystems to support even some of the most complex pedagogies with the help of audio-visuals and gaming effects [6]. However, it takes much more serious thinking to overcome several challenges while implementing educational technology. Several key factors affect the implementation of technological innovations in learning environments [5]. These include infrastructure preparedness, educator preparation, and cultural sensitivity as being integral to determining successful implementations in educational technology.

Over the past few years, artificial intelligence (AI) has proved to be a revolutionary innovation in educational technology [7]. AI technologies, given their capability to handle massive amounts of information, recognize patterns, and replicate certain human intellectual processes, and have opened never seen before possibilities for personalized learning experiences. The integration of AI into education systems marks more than an advancement of existing technologies; it marks a paradigm shift in the way knowledge is transferred, understood, and measured. This radical integration has the potential to address age-old challenges of educational access, personalization, and evolution of methods in numerous learning environments [7].

1.3 Artificial Intelligence and its Positive Impacts on Education

The deployment of AI in educational institutions has shown considerable potential to transform the teaching and learning processes at all levels of education. From pre-schooling to higher learning institutions, AI-supported technologies have been bringing about a revolution in traditional educational paradigms through offering flexible, customized, and accessible learning spaces. The following sections will explore various aspects of the role played by AI in education, highlighting its capacity to enhance accessibility, customization, and efficiency in offering educational services.

1.3.1 Personalized Learning

AI has become one of the most talked about topics in education in recent years. It holds a lot of promise to transform and improve many areas of the present education environment [8]. AI technologies, powered with machine learning algorithms and data analytics, open new opportunities for tailoring learning experiences, improving instructional strategies, and optimizing administrative processes. Applications of AI in the present education system are not limited to classrooms, personalized learning, assessment, various administrative tasks, and data analytics, but have the potential to revolutionize the whole educational experience for learners and educators.

The emergence of AI-based adaptive learning platforms has transformed the landscape of personalized education and ushered in an era of truly individualized learning experiences. The effectiveness of AI-powered personalization has been remarkable and significantly visible throughout the market. Intelligent tutoring systems, one of the pioneering applications of AI in education, provide individualized instruction, catering to each individual's needs and requirements, adapting to the learner's progress and providing immediate feedback.

Sophisticated AI algorithms can analyze intricate patterns in learner's performance, learning preferences, and progression rates with unprecedented precision [9–11]. These platforms adjust their learning content's difficulty levels dynamically with pacing and presentation methods, modifiable in real time, creating a responsive learning environment that adapts to each learner's unique needs. These platforms simultaneously support varied learning modes in unique ways and the effects of these modes on diverse learners have been remarkably positive [11]. Visual learners experience dynamic interactive graphics and augmented reality experiences, whereas auditory learners experience complex voice-based instruction and natural learning processing. Kinesthetic learners participate through immersive simulations and interactive exercises, thus enabling all learning styles to be supported [12]. AI-powered personalization is not just about adapting content, more advanced AI systems can sense the emotional state and stress levels of learners and tailor the learning experience accordingly. This is an emotional intelligence capability, which has helped reduce the anxiety levels of learners and improved their learning retention capabilities as well.

1.3.2 Inclusivity

AI-powered systems have become more precise in addressing specific learning challenges. Advanced natural language processing allows for real-time captioning with high accuracy, while computer vision systems allow for immediate text-to-speech conversion for visually impaired learners. These technologies have reduced learning barriers while improving learners' confidence and independence [8,13]. Recent breakthroughs in AI-supported learning exemplify how machine learning algorithms can predict and prevent problems in learning before they become major barriers. Such AI-based early-intervention systems can identify potential problems in learning with great accuracy, thus providing proactive support that reduces educational setbacks [14]. With the capabilities of AI, educational institutions can develop more interactive, responsive, and efficient learning environments that ultimately enhance educational outcomes and prepare learners for the demands of the future [15].

1.3.3 Global Access to Education

The deployment of AI-driven solutions has dramatically changed the access to education across the world in underserved communities. AI along with its capabilities of rapidly solving complicated problems within all areas of human life and simulating human intelligence, especially in educational functions, has resulted in making distance education an efficient method in education [9].

The role of AI in solving educator shortages has been especially important. AI-based virtual tutoring systems have not only ensured continuity of education in areas with acute educator shortages but also improved the learning outcomes of learners. Such systems have also reduced educational disparities between urban and rural areas [16]. AI-based solutions have transformed the way education is delivered in resource-constrained environments by optimizing resources. AI-based resource allocation systems have transformed the efficiency of institutions by cutting operational costs while improving the efficiency of content delivery [16]. Predictive analytics have also enabled better resource distribution, providing learners with an interactive environment that reflects cooperation and increases the accuracy of results in distance learning, thus making education accessible globally.

1.3.4 Efficiency in Teaching

The use of AI technologies in education has revolutionized the face of teaching efficiency, especially concerning the management of routine administrative tasks and assessment. AI systems can manage a great proportion of routine administrative tasks, thereby opening tremendous opportunities for more pedagogical focus [17]. This decrease in administrative burden has not only helped learners get more opportunities in upskilling themselves without depending on someone else, but has also helped in increased time spent on direct teaching activities, fundamentally impacting the quality of educator–learner interactions [18].

Beyond mere time saving, this administrative efficiency holds extensive implications, especially on education. Educational outcomes improve as educators report higher job satisfaction when they leverage AI-based administrative systems for teaching. Furthermore, educators' ability to pay more attention to each learner increases as these administrative systems reduce the amount of routine work performed, leading to lower educator burnout rates, thus enhancing retention and educational endurance.

AI tools have transformed the traditional assessment and feedback processes, making it more convenient for educators as well as learners. Analysis of learners' assignments across multiple disciplines shows that AI-assisted grading is remarkably consistent with expert human graders while significantly reducing assessment time. This increase in efficiency has profound implications on educators' satisfaction and quality of education.

The sophistication of AI-powered feedback systems goes beyond basic grading capabilities. Advanced AI systems are more capable of giving subtle, positive feedback that helps learners learn. Research has shown that AI-generated feedback is extremely advanced in analyzing complex aspects of a learner's work, including contextual and nuanced understanding in written assignments, pattern identification in learner misconceptions in multiple subjects, and the ability to give personalized suggestions for improvement that actually improves subsequent assignment performance [9]. These systems also create detailed progress reports that help educators identify and address learning gaps more effectively.

The integration of AI in assessments has particularly transformed teaching efficiency. Real-time AI assessment tools enable educators to adapt their instructional approaches dynamically. Educators using these tools show marked improvement in their ability to identify and address learners' learning needs promptly. This enhanced responsiveness has led to substantial improvements in learning outcomes and engagement levels of learners [13].

The evolution in AI has also changed teaching practices concerning collaborative teaching. The use of AI helps educators to collaborate more effectively and share more resources. In addition, AI-driven efficiency tools help to increase the professional development opportunities of educators. Educators who make use of AI-powered professional development tools can learn more technological skills while improving their pedagogical performance. These automated efficiency tools create time for critical professional growth and development for educators as well as learners.

1.4 Challenges of AI in Education

While AI promises to greatly improve learning experience and achievement, its application within education systems is loaded with serious challenges. Problems from ethics to equity and access issues raise serious questions about AI's use in education, and therefore, thoughtful consideration and progressive strategies are needed to address these challenges. The following sections highlight critical challenges that need to be addressed by educators, policymakers, and technology developers if AI is to be integrated in real life for good in the education system.

1.4.1 Ethical Concerns

The adoption of AI in education has introduced various significant ethical challenges that should be dealt with carefully and with a systemic approach. It is quite alarming that some learners have been extensively relying on AI tools, without making any efforts to learn [19]. They risk submitting assignments without even cross-checking the content, blindly copy pasting from generated texts from these AI tools, making it unauthenticated and lacking their actual academic ability. Thereby, a robust data protection framework in educational settings should emerge as an urgent imperative.

Apart from that, another challenge is of simple data protection, wherein these security issues have a much broader impact [19]. Educational AI system breaches could have lasting effects on learners' trust and engagement in the learning process. There has been a decline in learners who are willing to use learning tools that are AI-driven, just because of the risk of data breaches [20].

Another critical ethical challenge in educational AI systems is algorithmic bias. AI algorithms tend to amplify and perpetuate existing social and cultural biases [21]. AI assessment systems powered by machine learning algorithms have shown patterns of bias. Studies have reported that systems that are not properly calibrated score learners with discrepancies among different demographic groups [20].

1.4.2 Over-Reliance on AI

The growing reliance on AI-based learning tools has sparked much debate about the loss of critical thinking skills and cognitive independence. A heavy reliance on AI-based learning tools corresponds with a decline of independent problem-solving abilities [22]. More disturbing, however, is that this effect accumulates over time, with learners demonstrating increasingly diminished abilities to engage in complex reasoning without AI support. The psychological effects of dependency on AI are equally worrisome. Over-reliance on AI tools has resulted in a decrease in the confidence of learners in their own analytical capabilities and also reduced their willingness to tackle complex problems independently [9,22].

The reduction of human interaction in learning environments presents significant developmental challenges. Mainly AI-based learning environments affect the development of social skills and emotional intelligence in several ways [22]. The learners exhibit a weakening in their development of empathy, have less capacity for face-to-face communication with learners facing challenges in direct interpersonal interactions, and a deterioration in emotional recognition capabilities as compared to learners who follow traditional or hybrid learning methods.

1.4.3 Accessibility and the Digital Divide

The integration of AI in the education sector has brought existing technological inequalities to the forefront and, in many instances, worsened them. There are many educational institutions worldwide which do not have the minimum technology infrastructure needed for AI. The technology readiness gap between urban and rural institutions has also increased since AI-based educational tools have been introduced [9]. Learners from more wealthy backgrounds are much more likely to have access to advanced AI-powered educational tools. This has created what experts have termed the AI education gap, a phenomenon that threatens to amplify existing achievement gaps between socioeconomic groups [23].

The implementation of AI systems poses compound challenges in schools operating in economically disadvantageous areas [14]. These institutions experience a greater cost burden in implementing and maintaining AI systems, have lower access to technical expertise for the maintenance and optimization of the systems, and are less capable of offering educator training on using AI tools.

1.5 Balancing Innovation with Human Values: Responsible AI and Sustainable Education

The incorporation of AI in the educational field represents a careful collaboration between technological progress and the protection of our core human values of learning. This section looks at the crucial connection of ethical use of AI, alignment with sustainable development goals, and the increasingly collaborative dynamics between human instructors and AI technologies. Through analyzing these elements, we can further our knowledge about how to employ the capability of AI while learning experiences are protected as ethical, equitable, and ultimately human values driven.

1.5.1 Responsible AI Implementation

The establishment of robust ethical frameworks and comprehensive regulatory guidelines is the need of the hour and has become increasingly important as AI reshapes educational landscapes globally. Government frameworks along with practical continuous implementation will be the key to a transformational change worldwide. Enhancement in trust and a decline in ethical challenging cases are directly associated with clear, transparent AI governance frameworks in institutions because adoption is recorded higher at institutes that use clear and transparent approaches for their implementation [24].

Stakeholder involvement has emerged as a fundamental pillar in ensuring responsible AI implementation within educational settings. Organizations that actively involve all key stakeholders including educators, learners, parents, and administrators, in their AI governance processes to keep everybody on the same page, have achieved remarkably better results. Most successful governance models maintain a balance between structure and flexibility. These inclusive approaches have led to higher satisfaction rates and more effective implementation outcomes across all measures [25]. The impact of major stakeholder engagement goes beyond merely better satisfaction metrics; it fundamentally reforms the integration of AI technologies in the educational environment.

The role of educator training in the implementation of ethical AI has proven to be particularly crucial. Institutions that provide such extensive ethics training to their educators have seen fewer algorithmic bias incidents and have shown dramatically improved cultural sensitivity in their applications of AI [24,25]. The reduced bias incidents underscore the need for preparing educators for both the technical aspects of AI implementation and the ethical considerations and cultural implications of these technologies.

The new landscape in educational AI, which is rapidly changing, together with new challenges and opportunities for ethical implementation, requires careful and proactive planning to address emerging concerns about data privacy and algorithmic fairness in next-generation AI systems. To balance innovation with the realization of protecting learners' privacy and treating them fairly, these challenges can become even more difficult [19]. The major challenge that remains here is that financial sustainability to implement AI in education necessitates proper balancing of investment at its initial setup and the sustenance and upgradation over time, for which the technology budget may be reserved for such long-term AI sustainability in educational institutions as indicated by financial modeling along with experience. This needs to not only cover maintenance but also upgradation, as well as necessary training.

There is a possible complex landscape at the intersection of ethical considerations and sustainability requirements that educational institutions have to navigate with great care [21]. Success in this particular area would call for comprehensive consideration not only of the immediate implementation needs but also long-term viability, financial impact, and ongoing evolution of the ethical standards in AI education. Such comprehensive approaches to both ethics and sustainability will become increasingly important as AI continues its expansion in educational settings.

1.5.2 AI and the UN's Sustainable Development Goal 4

The integration of AI with global education goals has been remarkable in transforming educational access and quality around the world. Emerging technologies like AI and big data are not mentioned in the Sustainable Development Goals (SDGs); the 2030 Agenda for Sustainable Development still claims that ICTs, in general, can substantially increase progress and bridge the digital divide across communities, including education [26]. It is very significant in terms of addressing the needs of marginalized populations to see the transformation of access to education through AI-enhanced learning environments [27]. Indeed, such technological solutions have led to a rise in participation among previously underserved communities and transformed the educational access landscape fundamentally. This impact is beyond just participation but has reformed the social mobility of communities. AI-based learning platforms show remarkable differences in tangible educational outcomes. These improvements demonstrate practical applications of AI for the democratization of education, enabling more equitable access to learning opportunities in the different global communities [28].

1.5.3 The Evolving Role of Educators and Human–AI Collaboration

The integration of AI in education has resulted in a transformation in the role of educators, reshaping their daily responsibilities as well as their professional identity. This evolution goes far beyond simple technological adoption, representing a paradigm shift in how educators approach their profession and interact with learners [29]. This evolution allows more time for educators to be concerned about developing a learner's creative thinking and emotional intelligence, thereby giving importance to unique human capabilities in a digitized interconnected world [29].

The relationship between educators and technology has significantly evolved. Rather than considering AI as a threat to replace them, progressive educators embrace it as an empowerment tool that enhances their ability to teach [2]. This collaborative model has paved the way for more personalized learning at the learner's own pace. This perspective shift has resulted in more collaborative approaches where educators collaborate with AI platforms, with each bringing their unique strengths to the educational process.

This new model of education has rediscovered the importance of emotional intelligence and interpersonal skills. Educators can, therefore, cultivate deeper relationships with learners, creating supportive feelings and facilitating a friendly environment of learning by providing human elements of education, thereby making it even more indispensable [8]. With this shift, new roles focusing on specializations have also developed in teaching. There are AI integration specialists to guide educators through changes brought about by these new technologies, while others focus on creating new approaches of teaching using the potential of human insights and intellectual power blended with AI capabilities [11]. In that sense, these developments bring with them new career opportunities and further professional growth within the teaching sector.

This development in the AI field has seen the effectiveness of hybrid teaching models coming to the fore. That is, thoughtfully using AI assistance with traditional ways of teaching, these outcomes are better, with improvements reaching high, above either method used exclusively. This outcome underlines the need for a focus on human-centered education using effective AI capabilities to drive enhancements and support in teaching [29]. As the technology of AI continues to advance, the role of educators will likely continue to evolve. However, what will remain constant is the importance of human educators. The most successful models of education will be those that are able to balance technological innovation with the irreplaceable human elements of teaching to create learning environments that prepare learners for a rapidly changing world and develop their individual potential.

1.6 Conclusion

The evolution of educational AI is becoming increasingly complex, and it can only be navigated with caution as it embodies the best and worst for future generations. One promising frontier is emotional intelligence, and it is predicted that AI that could be emotionally aware and be able to improve learner engagement as it will enhance support systems for both educators and learners. This development suggests a future where educational technology can meet the full range of human learning needs for better performance.

With this, maintaining educational fairness becomes a major concern in the growingly AI-driven educational landscape. Current frameworks for equitable AI implementation show great promise, which could potentially reduce ethical challenges while promoting inclusive educational practices. Success in addressing these challenges while maximizing the benefits of AI integration will require continued innovation, careful attention to equity concerns, and ongoing commitment to maintaining the essential human elements of education.

References

[1] A. M. Al Samman, "Harnessing Potential: Meta-Analysis of AI Integration in Higher Education," in *2024 ASU International Conference in Emerging Technologies for Sustainability and Intelligent Systems, ICETSIS 2024*, Institute of Electrical and Electronics Engineers Inc., 2024, pp. 1656–1662. doi: 10.1109/ICETSIS61505.2024.10459420.

[2] N. Firdausi *et al.*, "Integrating Smart Education in Basic Education: A Systematic Review of Roles and Challenges," in *11th International Conference on ICT for Smart Society: Integrating Data and Artificial Intelligence for a Resilient and Sustainable Future Living, ICISS 2024 – Proceeding*, Institute of Electrical and Electronics Engineers Inc., 2024. doi: 10.1109/ICISS62896.2024.10751462.

[3] UNESCO "Out-of-School Children and Youth." UNESCO UIS. https://uis.unesco.org/en/topic/out-school-children-and-youth (Accessed Jan 12, 2025).

[4] S. Yang and H. Bai, "Research on School Education Reform Under the Background of '5G + Education,'" in *Proceedings – 2020 International Conference on Information Science and Education, ICISE-IE 2020*, Institute of Electrical and Electronics Engineers Inc., Dec. 2020, pp. 386–389. doi: 10.1109/ICISE51755.2020.00090.

[5] N. Gridchina, N. Savvina, and S. Zavyalov, "Topical Issues of the Use of Artificial Intelligence Technologies in Education and Law," in *Proceedings – 2023 3rd International Conference on Technology Enhanced Learning in Higher Education, TELE 2023*, Institute of Electrical and Electronics Engineers Inc., 2023, pp. 8–11. doi: 10.1109/TELE58910.2023.10184378.

[6] B. Li, "Modern Education Technologies in Higher Special Education," in *Proceedings of the 1st International Workshop on Education Technology and Computer Science, ETCS 2009*, Institute of Electrical and Electronics Engineers Inc., 2009, pp. 622–625. doi: 10.1109/ETCS.2009.672.

[7] L. Chen, P. Chen, and Z. Lin, "Artificial Intelligence in Education: A Review," *IEEE Access*, vol. 8, pp. 75264–75278, 2020. doi: 10.1109/ACCESS.2020.2988510.

[8] D. Kumar, A. Haque, K. Mishra, F. Islam, B. Kumar Mishra, and S. Ahmad, "Exploring the Transformative Role of Artificial Intelligence and Metaverse in Education: A Comprehensive Review," *Metaverse Basic and Applied Research*, vol. 2, p. 55, Dec. 2023. doi: 10.56294/mr202355.

[9] S. Khedrane, "The Role of Artificial Intelligence in Distance Education", *International Journal of Social Communication*, vol. 11, no. 1, pp. 72–83, 2024. doi: 10.53284/2120-011-001-006.

[10] L. Guo and B. Wen, "Analysis and Reflections on the Use of Artificial Intelligence in Teaching Management and Evaluation–Analysis Based on CiteSpace," in *Proceedings of the IEEE International Conference on Software Engineering and Service Sciences, ICSESS*, IEEE Computer Society, 2023, pp. 206–211. doi: 10.1109/ICSESS58500.2023.10293125.
[11] M. A. Rind, M. A. Al Qudah, and P. Aliyev, "Determining the Impact Artificial Intelligence on Development of Higher Education," in *2024 ASU International Conference in Emerging Technologies for Sustainability and Intelligent Systems, ICETSIS 2024*, Institute of Electrical and Electronics Engineers Inc., 2024, pp. 1486–1489. doi: 10.1109/ICETSIS61505.2024.10459626.
[12] I. Vostroknutov, S. Grigoriev, and L. Surat, "Search for a New Paradigm of Education and Artificial Intelligence. Place and Role of Artificial Intelligence in the New Education System," in *Proceedings – 2021 1st International Conference on Technology Enhanced Learning in Higher Education, TELE 2021*, Institute of Electrical and Electronics Engineers Inc., Jun. 2021, pp. 80–82. doi: 10.1109/TELE52840.2021.9482486.
[13] T. Baksa, M. Konecki, and M. Konecki, "High School Students' Perception of AI and Its Future Impact on Education," in *2024 12th International Conference on Information and Education Technology, ICIET 2024*, Institute of Electrical and Electronics Engineers Inc., 2024, pp. 215–219. doi: 10.1109/ICIET60671.2024.10542754.
[14] O. Hurenko, I. Bohdanov, N. Tsybuliak, H. Lopatina, Y. Suchikova, and A. Popova, "Development of an Inclusive Educational Environment in Higher Education Institutions: A Project Approach Using IDEF0," in *International Scientific and Technical Conference on Computer Sciences and Information Technologies*, Institute of Electrical and Electronics Engineers Inc., 2023. doi: 10.1109/CSIT61576.2023.10324022.
[15] S. Gupta, R. R. Dharamshi, and V. Kakde, "An Impactful and Revolutionized Educational Ecosystem using Generative AI to Assist and Assess the Teaching and Learning benefits, Fostering the Post-Pandemic Requirements," in *2nd International Conference on Emerging Trends in Information Technology and Engineering, ic-ETITE 2024*, Institute of Electrical and Electronics Engineers Inc., 2024. doi: 10.1109/ic-ETITE58242.2024.10493370.
[16] K. Yunusov, B. Berdiyev, and B. Jovliev, "A Well Structured System for Learning and also Evaluating the Outcome for the Better Learning using Artificial Intelligence in Higher Pedagogy," in *2024 4th International Conference on Advance Computing and Innovative Technologies in Engineering, ICACITE 2024*, Institute of Electrical and Electronics Engineers Inc., 2024, pp. 857–861. doi: 10.1109/ICACITE60783.2024.10617101.
[17] M. Mukhammadsidiqov, S. Akmalov, and K. Akhmedov, "An Intensive Review and Assessment of Impacts of AI in the Field of Teaching(Edu)," in *2024 4th International Conference on Advance Computing and Innovative Technologies in Engineering, ICACITE 2024*, Institute of Electrical and Electronics Engineers Inc., 2024, pp. 1013–1017. doi: 10.1109/ICACITE60783.2024.10616640.
[18] R. Patel, P. Bajaj, A. Kumar, A. Kumari, V. Rai, and S. Kumar, "ChatGPT in the Classroom: A Comprehensive Review of the Impact of ChatGPT on Modern Education," in *ISED 2023 – International Conference on Intelligent Systems and Embedded Design*, Institute of Electrical and Electronics Engineers Inc., 2023. doi: 10.1109/ISED59382.2023.10444568.

[19] M. Zeer, R. W. Siaj, J. A. Ghannam, and M. Kanan, "Ethics of Artificial Intelligence in University Education," in *2nd International Engineering Conference on Electrical, Energy, and Artificial Intelligence, EICEEAI 2023*, Institute of Electrical and Electronics Engineers Inc., 2023. doi: 10.1109/EICEEAI60672.2023.10590285.

[20] R. Sharma, P. Kumar, D. K. Singh, D. Suri, P. Rajput, and S. Kumar, "The Intersection of AI, Ethics, and Education: A Bibliometric Analysis," in *ICCDS 2024 – International Conference on Computing and Data Science*, Institute of Electrical and Electronics Engineers Inc., 2024. doi: 10.1109/ICCDS60734.2024.10560363.

[21] J. R. M. Montezuma and M. Chong, "Generative Artificial Intelligence Impact on Education and Industry: An Ethical Dimension," in *EDUNINE 2024 - 8th IEEE World Engineering Education Conference: Empowering Engineering Education: Breaking Barriers through Research and Innovation, Proceedings*, Institute of Electrical and Electronics Engineers Inc., 2024. doi: 10.1109/EDUNINE60625.2024.10500608.

[22] M. Biswas and J. Murray, "The Impact of Education Level on AI Reliance, Habit Formation, and Usage," in *ICAC 2024 - 29th International Conference on Automation and Computing*, Institute of Electrical and Electronics Engineers Inc., 2024. doi: 10.1109/ICAC61394.2024.10718860.

[23] H. Li, "AI in Education: Bridging the Divide or Widening the Gap? Exploring Equity, Opportunities, and Challenges in the Digital Age," *Proceedings -The 2023 International Conference on Education, Philosophy, and Social Sciences, ICEPSS 2023*, vol. 8, no. 1, pp. 355–355, 2023. doi: 10.56028/aehssr.8.1.355.2023.

[24] D. Shao, N. Marwa, and D. Matendo, "Regulatory Strategies for Fostering Responsible AI Innovation in African Economies," in *1st International Conference on the Advancements of Artificial Intelligence in African Context, AAIAC 2023*, Institute of Electrical and Electronics Engineers Inc., 2023. doi: 10.1109/AAIAC60008.2023.10465404.

[25] C. Sanderson, Q. Lu, D. Douglas, X. Xu, L. Zhu, and J. Whittle, "Towards Implementing Responsible AI," in *Proceedings – 2022 IEEE International Conference on Big Data, Big Data 2022*, Institute of Electrical and Electronics Engineers Inc., 2022, pp. 5076–5081. doi: 10.1109/BigData55660.2022.10021121.

[26] R. Walshe, K. Casey, J. Kernan, and D. Fitzpatrick, "AI and Big Data Standardization: Contributing to United Nations Sustainable Development Goals," *Journal of ICT Standardization*, vol. 8, no. 2, pp. 77–106, Apr. 2020, doi: 10.13052/jicts2245-800X.821.

[27] K. Yilma, "Emerging Technologies and Human Rights at the United Nations," *IEEE Technology and Society Magazine*, vol. 42, no. 1, pp. 54–64, Mar. 2023, doi: 10.1109/MTS.2023.3241297.

[28] R. Wang, C. Li, X. Li, R. Deng, and Z. Dong, "GenAI4Sustainability: GPT and Its Potentials For Achieving UN's Sustainable Development Goals," *IEEE/CAA Journal of Automatica Sinica*, vol. 10, no. 12, pp. 2179–2182, Dec. 2023. doi: 10.1109/JAS.2023.123999.

[29] R. Zhang, Y. Qiu, and Y. Li, "An Empirical Study on Human-Machine Collaborative MOOC Learning Interaction Empowered by Generative AI," in *Proceedings – 2023 International Symposium on Educational Technology, ISET 2023*, Institute of Electrical and Electronics Engineers Inc., 2023, pp. 116–120. doi: 10.1109/ISET58841.2023.00031.

2

Leveraging AI to Foster Inclusive Quality Education in Higher Educational Institutes

Rahul Sagwal, Abhinav Sharma and Janakarajan Ramkumar

2.1 Introduction

Artificial intelligence (AI) has brought about a revolution across various sectors, and education is no exception. The field of AI is rapidly expanding and is transitioning the teaching–learning practices in higher education (HE). AI can be leveraged in higher educational institutes to make the learning practices more inclusive, adaptive, and equitable. Traditional teaching techniques often struggle to cater to the diverse learning needs of students coming from various social backgrounds, having different learning styles and abilities. However, AI-powered teaching–learning techniques offer solutions to these gaps by providing personalized learning experiences, real-time feedback, and accessibility for all.

AI's contribution towards higher education has been on a rising trend in recent months and is likely to continue to rise exponentially in the near future too. Ranging from automated content generation through generative AI, ChatGPT, and intelligent individual tutoring systems to tools that enhance accessibility of higher education to students with learning disabilities, AI is playing a crucial role. This chapter explores the transformative potential of AI in fostering inclusive learning within higher education. By leveraging advanced tools such as deep learning, large language models, text to speech, real-time language translation, and predictive analytics, engineering institutes can better identify and address the individual learner needs of diverse students and thus, will be able to ensure that no learner is left behind. With the help of data-driven insights gathered through these AI tools, teachers will be able to make informed decisions on individual student's performances and can modify the teaching strategies and content in real time for the needful students. This chapter first identifies the issues faced by students with learning disabilities in higher education. Then it highlights

DOI: 10.1201/9781003560814-2

the transformative potential of AI-powered tools in the field of education, followed by discussions on current applications, emerging trends, and ethical considerations of using AI in higher education. The chapter aims to provide a comprehensive framework for leveraging AI-driven technologies to empower diverse learners and build more equitable learning ecosystems for all.

2.2 AI in Higher Education

AI in Education (AIEd) can be looked upon from different perspectives: (1) learner oriented, (2) instructor oriented, and (3) institute oriented. To set the pretext for our consideration of AI for higher education, a summarization of these categorizations has been provided in Figure 2.1 (Shamsuddinova, Heryani, and Naval 2024). In addition, institute-oriented AIEd includes: student selection during admissions, course planning and restructuring, timetabling, dropout and students-at-risk identification, and e-proctoring (Holmes and Tuomi 2022).

Here, it is worth noting that intelligent tutoring systems (ITS) are one of the most adopted AI tools in higher education. They allow 1:1 attention to the individual student's problems and provide step-by-step guidance for enhanced understanding of complex subjects, including advanced mathematics and computational languages. A recent study in AIEd (Shamsuddinova, Heryani, and Naval 2024) conducted university educator interviews to establish 20 codes for their responses about their perception of AI in education and mapped these with four broad themes (see Figure 2.2). The major focus of the current work is limited to learner-oriented AIEd, which is also primarily tailored to the needs of learners with disabilities.

ITS opens up the spectrum of AI-based learning by bringing in both cognitive and non-cognitive factors affecting the learning process. It is difficult to analytically comprehend the cognitive parameters and include subject-wise variability. As compared to conventional tutoring, ITS has an advantage as it provides real-time strategies for all stakeholders including students, tutors, and managers. With a better understanding of the entire system this helps in fostering a culture of collaborative learning.

ITS enables information-rich pedagogical methods and enriches policy makers with substantial tailored content for making learning skills in students simpler to enable.

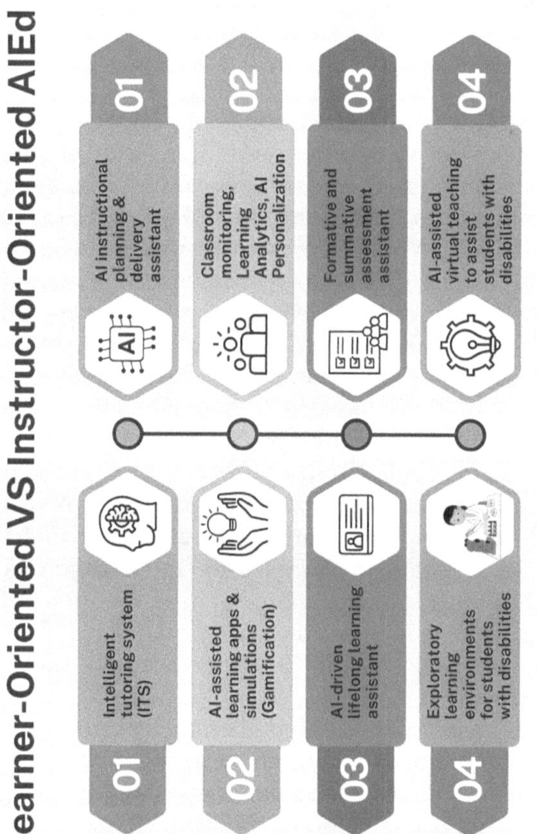

FIGURE 2.1
Taxonomy for AI in education. (Authors based on Shamsuddinova, Heryani, and Naval, 2024.)

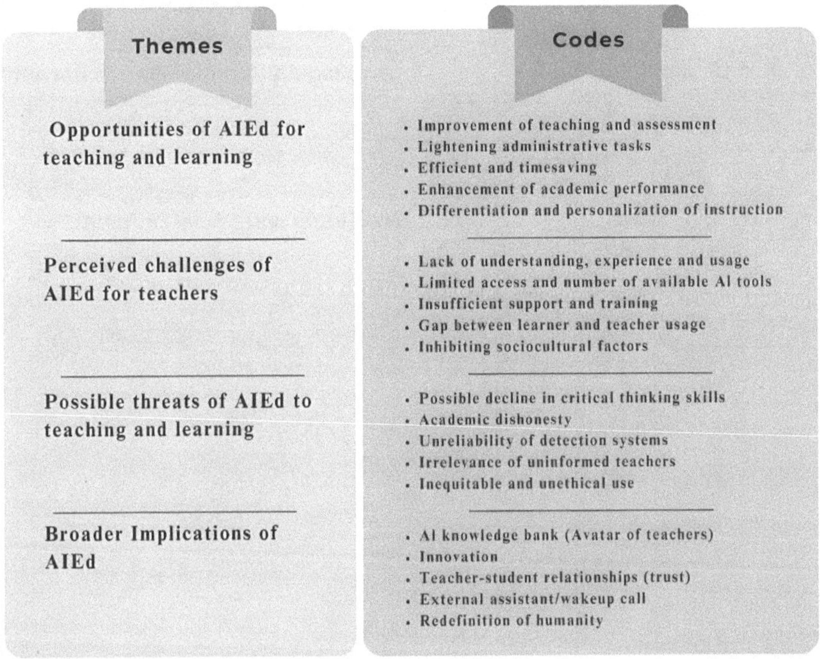

FIGURE 2.2
Correlation of AIEd themes with codes – educator's interview responses. (Authors based on Shamsuddinova, Heryani, and Naval, 2024.)

2.3 Learning Difficulties

Students coming from diverse backgrounds and cognitive abilities demonstrate different learning speeds (Tielman et al. 2024). The major reasons behind the learning difficulties faced by learners are summarized in Figure 2.3.

2.4 Use of AI to Address these Difficulties

When it comes to the usage of AI to support learners with learning disabilities, AI finds significance right at the start. It is leveraged for identification of students with special needs in learning and the type of associated learning needs also (El Naggar, Gaad, and Inocencio 2024). Bressane et al. (2024)

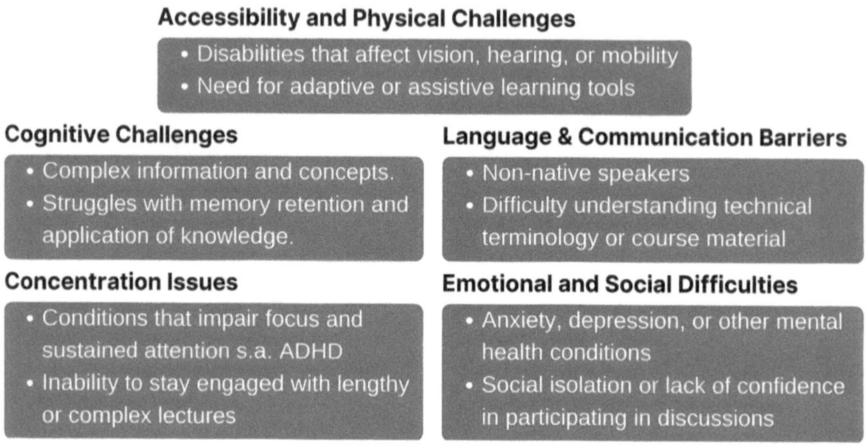

FIGURE 2.3
Learning difficulties. (Authors.)

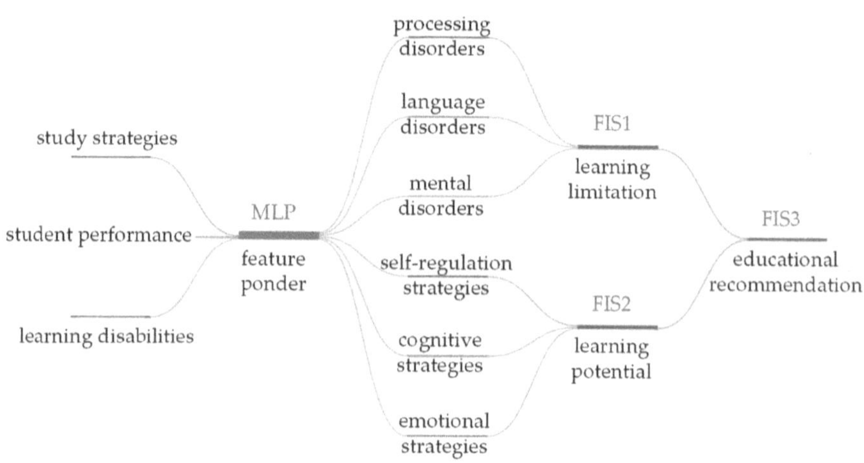

FIGURE 2.4
AI-based decision support system. (Bressane et al. 2024.)

present an interesting way of using AI, particularly a Fuzzy Inference System (FIS)-based Decision Support System (DSS), to make recommendations about the educational strategy that is best for the individual based on their study strategy and learning disabilities (see Figure 2.4). In this way, it becomes easier for the educator and learners to formulate groups in the classroom so that students with similar types of educational strategy can work best with their peers for an enhanced learning environment. This study also highlights how much impact different types of learning disabilities have on the education of

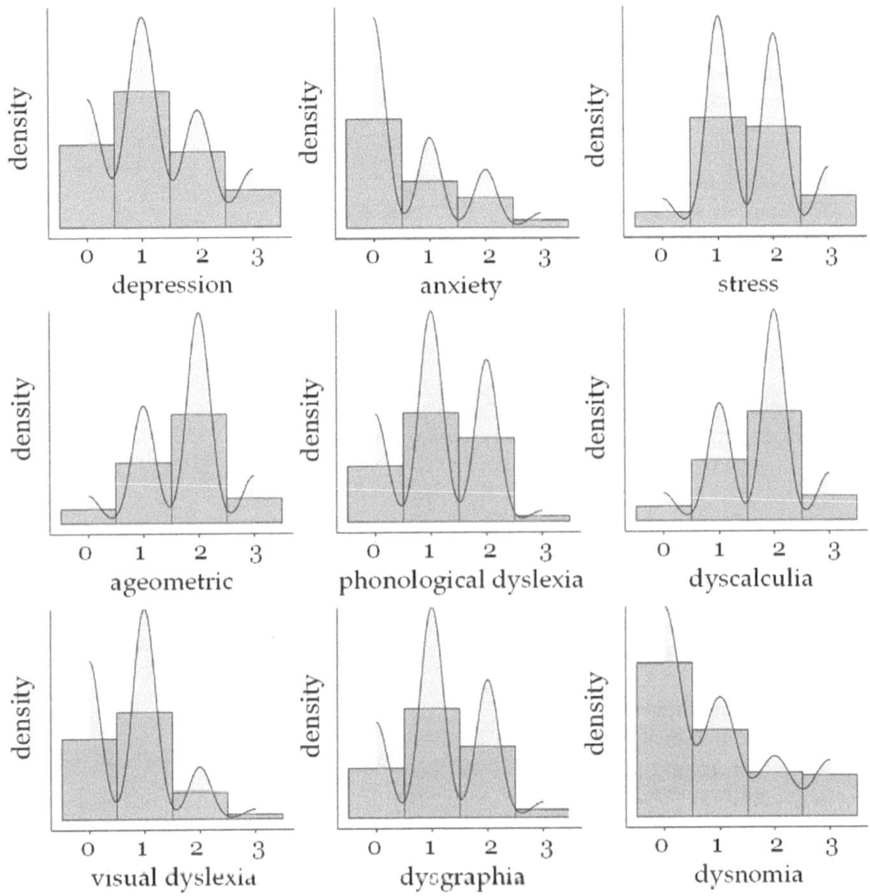

FIGURE 2.5
Occurrence of learning disabilities: 0 – never, 1 – rarely, 2 – often, and 3 – always. (Bressane et al. 2024.)

the students (Conde and Rodríguez-Sedano 2024; see Figure 2.5). These are then joined with study strategies adopted by various students to decide their significance (see Figure 2.6), before passing these on to the AI-powered DSS framework.

Figure 2.6 represents the coefficient of determination in decreasing order and associated weights to indicate the most impactful factors for the academic performance of students. As an output of this framework, individual students get recommendations of the best study strategy that suits their needs. These recommended educational strategies can be seen in Figure 2.7. These individualized strategies foster inclusivity through AI-driven adaptive approaches according to the student's learning profile (Bressane et al. 2024).

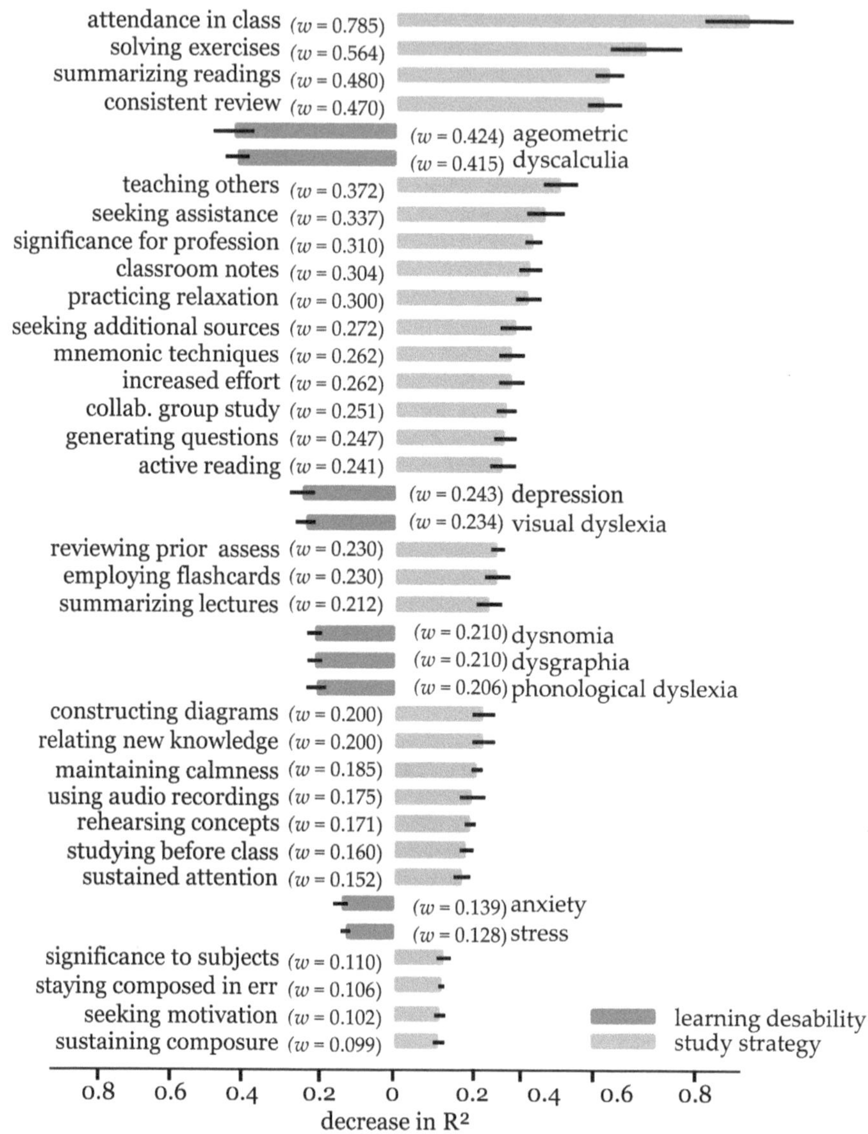

FIGURE 2.6
Coefficient of determination for different scenarios. (Bressane et al. 2024.)

Based on the principles of Universal Design for Learning (UDL), primarily the what, why, and how of learning, Song et al. (2024) have proposed a framework for educators and learners to design inclusive learning opportunities tailored to AI education (see Figure 2.8). This article demonstrates AI instruction with six practices with pedagogical examples.

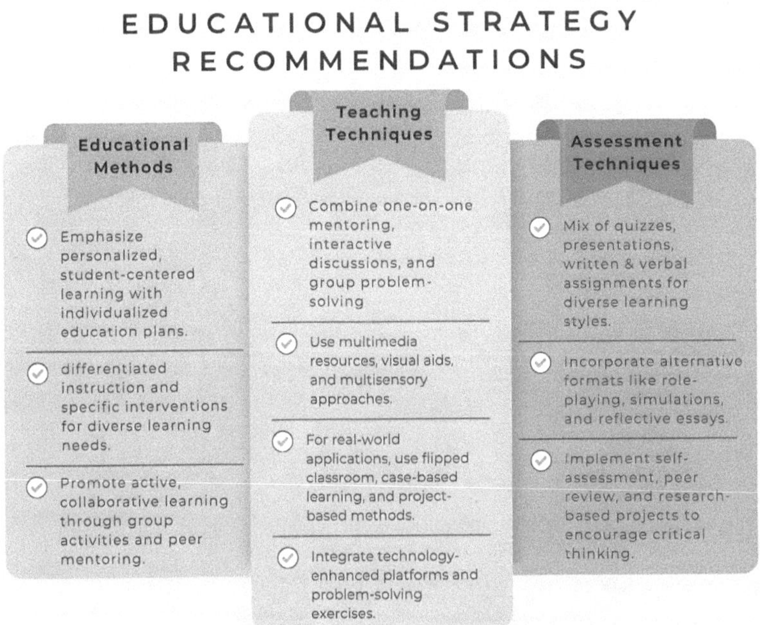

FIGURE 2.7
Educational strategy recommendations. (Authors, based on Bressane et al. 2024.)

2.5 Modern AI Tools for AI in Higher Education

Traditional teaching methods employ a uniform teaching–learning pace, fixed curriculum, and universal learning strategy for all, which might not address individual student's learning needs. On the contrary, AI-powered teaching–learning tools allow personalization, real-time feedback to tutors and students enabling them to quickly adapt, and supports scalability – large classrooms – without compromising individual attention. Some of these AI tools used in higher educational institutes are depicted in Figure 2.9 (Lee et al. 2024; Moylan, Code, and O'Brien 2025).

Intelligent Tutoring Systems (ITS): ITS, which are predominantly computer software, are used to enhance the learning process. These uses AI algorithms to analyse student's performance and learning curves and adapt the difficulty level and upcoming learning content based on real-time feedback. ITS incorporate knowledge-tracking models to predict individual performance and continuously update personalized learning paths. One such system, ALEKS (Assessment and Learning in Knowledge Spaces), has been used (Puri et al. 2024) to understand gaps in the prerequisite knowledge of students.

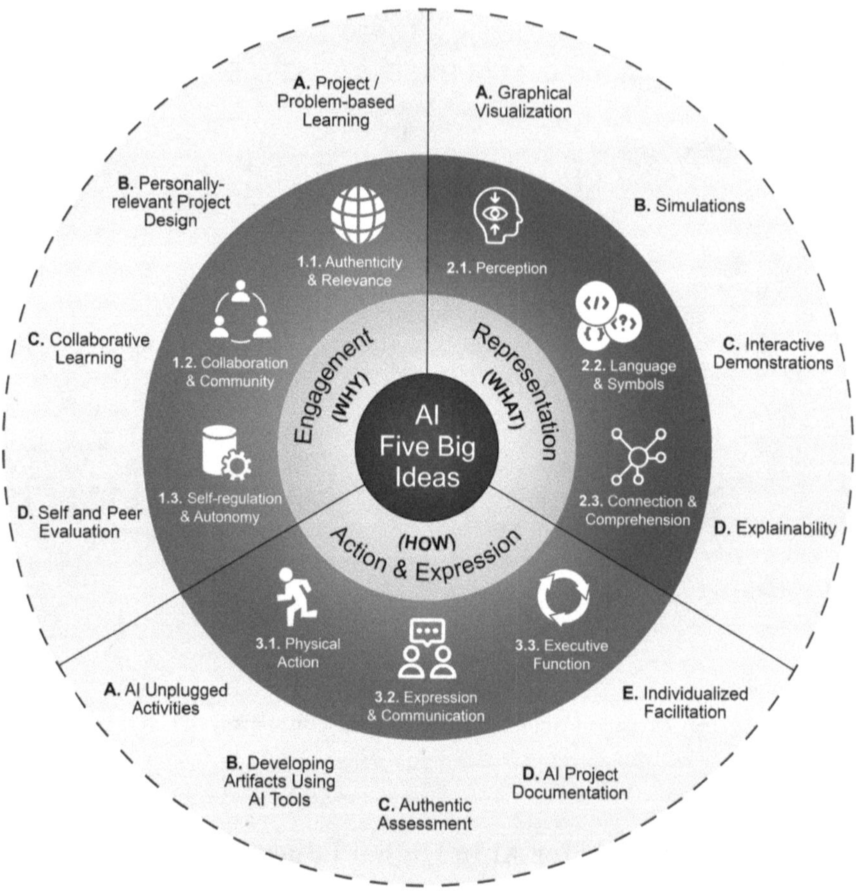

FIGURE 2.8
Framework for inclusive AI learning design for diverse learners. (Song et al. 2024.)

Assistive Technologies: These assist students with learning difficulties through tools such as text-to-speech, speech-to-text, language translation, and loud screen readers, etc. Furthermore, specialized AI companions or chatbots provide emotional support too. Examples include Microsoft immersive reader and Google translator.

Virtual and Augmented Reality (VR/AR): These provide an immersive learning experience to students and offer better visualization of complex concepts. Virtual field trips and collaborative learning irrespective of geographical location are additional benefits of AR/VR technologies. Examples include ClassVR, Google Expeditions, etc.

Speech Recognition and Natural Language Processing (NLP): Google Assistant, Amazon Alexa, and ChatGPT powered bots are many such examples that effectively utilize speech recognition technology and aid

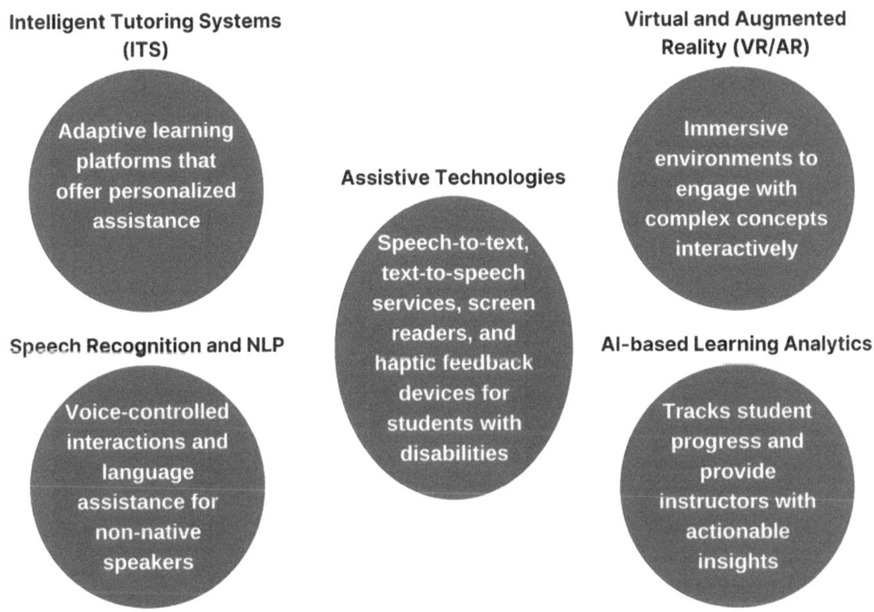

FIGURE 2.9
AI tools in higher education. (Authors.)

in the pronunciation and conversational practice process of students with difficulties.

AI-based Learning Analytics: AI algorithms offer automatic and real-time assessments of students that helps the tutor in identifying students at-risk and provides early intervention through personalized study material, study plans, and assignment recommendations. This can further help in curriculum optimization. Examples include IBM Watson Education.

2.6 Ethical Implications of AI in Higher Education

The transition from traditional learning practices to the realm of AI demands careful planning, as these tools if used unethically can severely impact the main purpose of learning in students (Alsarawi 2024). Furthermore, when it comes to utilizing AI to enhance the learning environment for students with learning disabilities it is important to consider the following ethical points:

- Digital Divide – Inclusion of AI in higher education is expected to make the ecosystem more equitable for students with learning disabilities but limited access to AI solutions can disproportionately affect such students, exacerbating inequities.

- Data Privacy and Consent – AI systems handling data about disabilities need stringent steps to protect privacy and avoid the misuse of such data. Additionally, in many cases of students with disabilities, it might become difficult for them to understand the informed consent process and requiring a dedicated, simplified, accessible consent process design.
- Algorithmic Bias – As the AI-driven processes are extremely dependent on data used to train the models, unintended exclusion might happen. Bias in training algorithms, if undetected, can result in tools that fail to meet the needs of students with different sensory, physical, or cognitive abilities.
- Over-personalization – AI tools are likely to make the learning environment more inclusive, yet they face challenges such as over-personalization. This might lead to the isolation of certain students instead of fostering an inclusive environment.
- Academic Integrity – AI systems must be designed for fair evaluation of students with disabilities considering their unique challenges. Cheating concerns also needs to be addressed.
- Dehumanization of Support – Overreliance on AI might lead to reduced interactions with faculty and peers. This might impact emotional and social support; a balance needs to be maintained in automation with necessary human interventions.
- Training for Educators – Developers should consider Universal Design Principles while designing and must make provisions to train the educators to avoid post-deployment barriers.
- Legal and Policy Frameworks – AI tools specially designed for students with disabilities must comply with legal and medical standards, wherever necessary.

2.7 Conclusions

AI has been at centre of all modern innovations. This very fact excites the application of AI within the pedagogical framework. Holistic education in technology-rich times cannot be thought of without the use of AI in policy making, teaching methods, evaluation systems, and the overall learning cycle.

This chapter underscores the importance of the integration of AI with higher education which holds transformative potential to address the needs of students facing learning disabilities and to make the education ecosystem more inclusive. Furthermore, we dive into the associated tools and methods for AI-based learning. Institutions can leverage AI tools to identify the barriers faced by students and can better cater to individual needs to provide them with an enhanced learning experience. AI-powered tools offer more personalization of learning strategies and learning pace, make education more accessible for all, and foster an equitable learning environment. However, ethical considerations are crucial when it comes to training AI models with huge amounts of data of students with learning disabilities to avoid its misuse. In addition, several other ethical implications such as algorithmic fairness, over-personalization, academic integrity, and human-centric learning, etc. need to be addressed before transitioning entirely to automated learning ecosystems. We have also tried to explore the use of ethical methods such that AI-based systems do not create an environment that lacks human effort and ethics.

References

Alsarawi, Aeshah A. 2024. "Managing and Learning from the Practicum Challenges of Graduate Entry Pre-Service Teachers of Students with Learning Disabilities: A Case Study." *Teaching and Teacher Education* 145 (July):104636. https://doi.org/10.1016/j.tate.2024.104636.

Bressane, Adriano, Daniel Zwirn, Alexei Essiptchouk, Antônio Carlos Varela Saraiva, Fernando Luiz De Campos Carvalho, Jorge Kennety Silva Formiga, Líliam César De Castro Medeiros, and Rogério Galante Negri. 2024. "Understanding the Role of Study Strategies and Learning Disabilities on Student Academic Performance to Enhance Educational Approaches: A Proposal Using Artificial Intelligence." *Computers and Education: Artificial Intelligence* 6 (June):100196. https://doi.org/10.1016/j.caeai.2023.100196.

Conde, Miguel Á., and Francisco J. Rodríguez-Sedano. 2024. "Is Learning Analytics Applicable and Applied to Education of Students with Intellectual/Developmental Disabilities? A Systematic Literature Review." *Computers in Human Behavior* 155 (June):108184. https://doi.org/10.1016/j.chb.2024.108184.

El Naggar, Alia, Eman Gaad, and Shannaiah Aubrey Mae Inocencio. 2024. "Enhancing Inclusive Education in the UAE: Integrating AI for Diverse Learning Needs." *Research in Developmental Disabilities* 147 (April):104685. https://doi.org/10.1016/j.ridd.2024.104685.

Holmes, Wayne, and Ilkka Tuomi. 2022. "State of the Art and Practice in AI in Education." *European Journal of Education* 57 (4):542–70. https://doi.org/10.1111/ejed.12533.

Lee, Daniel, Matthew Arnold, Amit Srivastava, Katrina Plastow, Peter Strelan, Florian Ploeckl, Dimitra Lekkas, and Edward Palmer. 2024. "The Impact of Generative AI on Higher Education Learning and Teaching: A Study of Educators' Perspectives." *Computers and Education: Artificial Intelligence* 6 (June):100221. https://doi.org/10.1016/j.caeai.2024.100221.

Moylan, Rachel, Jillianne Code, and Heather O'Brien. 2025. "Teaching and AI in the Postdigital Age: Learning from Teachers' Perspectives." *Teaching and Teacher Education* 153 (January):104851. https://doi.org/10.1016/j.tate.2024.104851.

Puri, Ashana, Stacy Brown, Michele Williams, and David Stewart. 2024. "Using ALEKS to Remediate Gaps in Pre-Requisite Knowledge for PharmD Students." *American Journal of Pharmaceutical Education* 88 (9): 101195. https://doi.org/10.1016/j.ajpe.2024.101195.

Shamsuddinova, Shakhnoza, Poonam Heryani, and Mark Anthony Naval. 2024. "Evolution to Revolution: Critical Exploration of Educators' Perceptions of the Impact of Artificial Intelligence (AI) on the Teaching and Learning Process in the GCC Region." *International Journal of Educational Research* 125:102326. https://doi.org/10.1016/j.ijer.2024.102326.

Song, Yukyeong, Lauren R. Weisberg, Shan Zhang, Xiaoyi Tian, Kristy Elizabeth Boyer, and Maya Israel. 2024. "A Framework for Inclusive AI Learning Design for Diverse Learners." *Computers and Education: Artificial Intelligence* 6 (June):100212. https://doi.org/10.1016/j.caeai.2024.100212.

Tielman, Myrthe L., Mari Carmen Suárez-Figueroa, Arne Jönsson, Mark A. Neerincx, and Luciano Cavalcante Siebert. 2024. "Explainable AI for All – A Roadmap for Inclusive XAI for People with Cognitive Disabilities." *Technology in Society* 79 (December):102685. https://doi.org/10.1016/j.techsoc.2024.102685.

3

AI and Inclusive Education: Enhancing Equity, Engagement and Excellence

Neha Kumari and Bhairab Chandra Patra

3.1 Introduction

Inclusive education represents the paradigm where every individual has an equal opportunity to learn. Such a setting values diversity and provides every learner with an avenue for smooth growth. Artificial intelligence (AI) has unlocked transformative tools that can bolster the impacts of inclusive education in handling all of students' divergent learning needs in an individualized and adaptive way (Jafarnia et al., 2023). AI brings about personalized learning pathways that adjust to the needs, abilities, or pace of every learner. This system is different from the older one-size-fits-all approach, where AI-based platforms emphasize looking into data about the student's progress, areas of strength, and weaknesses as they change the learning content to better match the student's needs (Srinivasa et al., 2022). This is especially beneficial to inclusive classrooms where students may have different cognitive, physical, or emotional needs. For example, AI-powered learning platforms, such as intelligent tutoring systems, can determine how well a student understands a subject and then make recommendations tailored to that student's needs. Adaptive learning tools might pose difficulties for advanced learners while at the same time offering increased support to learners who need more assistance (Woolf, 2010). Speech recognition software helps students with speech and language impairments by allowing them to interact with education content by voice commands where input is more of a barrier than usual.

AI, and other assistive technologies, can cater to the unique needs of students with disabilities and make learning material more easily accessible to learners. They offer educators an opportunity to meet the challenges of accommodating learners with disabilities in physical, sensory, or cognitive ways in one classroom (Patel et al., 2022). AI has helped to enable highly powerful robotics, voice-command programs, and eye-tracking devices, among others, which allow physically impaired students to access learning content in a better way. For instance, an extremely mobility-disabled student

DOI: 10.1201/9781003560814-3

can utilize AI-based assistive tools to seek digital content independently and submit assignments. Text-to-speech (TTS) and speech-to-text (STT) systems aid any kind of reading or writing disability for students by converting a written text to spoken words and vice versa (Almgren Bäck et al., 2024). This can help students with dyslexia, for example, because such students are likely to find reading text a problem, but they can still learn through listening. Augmentative and alternative communication (AAC) devices and AI-based communication supports enable students who find it difficult to communicate to be able to give voice to their thoughts and participate in classwork (Griffiths et al., 2024). AI can analyze humungous datasets to furnish educators with deep insights into how students are learning as well as their needs. This allows teachers to develop more effective, inclusive curricula and improve their methods of teaching. Data-driven insights help teachers identify areas of learning a particular student or group of students needs and administer interventions early, ensuring they do not get too far behind. They can track the progress of individual students or groups at any point in time, monitor the journey of learning and in turn make data-driven adjustments to lesson plans or to instructional methods (Bambrick-Santoyo,2010), and determine student outcomes with historical data, which may help guide teachers to strategies that can allocate resources and attention based on where it is most needed.

AI can target the unique needs of the learner, while at the same time fostering an open environment where every student has collaborative participation regardless of their personal background or abilities. AI-advanced collaborative learning spaces can facilitate matching students at different levels to work in teams, and then there will be peer-assisted learning. This can break down the barriers that society puts forward while promoting inclusiveness by letting students befriend and support each other. AI-powered translation tools for languages make it easier for students who communicate in different languages to interact with education material and with one another (Alharbi, 2023). This allows the learning environment to be broader as students learn in their mother tongue but gradually change over to a second language. Thus, AI can look out for facial expressions, tone of voice, and body language to notice where a student seems disinterested or frustrated. Educators are generally able to intervene just in time to make sure that emotional and social reasons do not become obstacles to learning.

AI enables teachers to develop more inclusive curricula with respect to diverse abilities, learning styles, and preferences of students. With AI, teachers will be able to design differentiated content that might be customized for the complexity or reading level and learning style preferred by each student. This makes it much easier to meet the needs of visual, auditory, kinesthetic, and verbal learners. Furthermore, AI-powered virtual or augmented reality can make learning more captivating for those students who somehow have difficulty with the usual modes of teaching through gamification. For instance, simulation environments might clarify abstract concepts for students with autism spectrum disorders or learning disabilities.

AI is also changing the way assessment takes place, from standardized testing to more dynamic assessment methods that are adaptive to an individual. AI-powered assessment systems can provide ongoing, formative assessments to measure student learning over a period of time rather than at predetermined fixed points, allowing better tracking of the progress of student learning (Alharbi, 2023). They provide alternative methods of assessment for students who may face barriers in accessing traditional testing formats because of disabilities. Some examples include that a test could be administered orally, meaning speech recognition software is allowed for students who have problems with dysgraphia, having difficulty writing.

3.2 Literature Review

Research in recent years on artificial intelligence applications in education, especially toward inclusive education, has been enormous. Inclusive education is supportive learning for all students, irrespective of one's ability, social class, economic background, or learning difficulties. Based on recent studies and emerging trends, this literature review highlights the theoretical framework, technological advancement, and practical application of AI in the context of inclusive education. Mainstreaming is based on the principle that all children, irrespective of their abilities in relation to physical, cognitive, or social-emotional development, have a right to be educated within mainstream settings (Ainscow, 2020). This approach emphasizes the removal of barriers to participation and learning in schools, calling for schooling systems to become more inclusive. Inclusion principles have been embraced by many, while their implementation can be rather difficult due to factors such as adherence to modifications according to different curricula in a school to ensure that students from different backgrounds get the right education and teachers cannot easily undertake sufficient differentiation in teaching. Teachers encounter various difficulties in their quest to attend to the individual needs of their learners of different abilities and backgrounds. Schools might lack the technological and human resources needed to ensure the successful implementation of inclusive education (Ngalim, 2019). Authentic methods of assessment may not consider the diversity of abilities of students within an inclusive learning setting.

Artificial intelligence will be the transformative agent for inclusive education. Personalized learning will be provided based on each student's needs. Through AI, adaptive learning systems can monitor the pupil's real-time progress and adjust the content for learning. Adaptive learning systems are designed to obtain information about student performance and their behavior and engagement so that appropriate paths of learning might be offered to every student, and thus, ensuring that effective learning becomes a reality for all (Halkiopoulos & Gkintoni, 2024). Some research studies depict how adaptive learning technologies can be beneficial in fostering inclusivity. For instance, AI-based ITS such as Carnegie Learning's "Cognitive Tutor" have been successful in showing a good outcome for the betterment of students' learning experiences of every type irrespective of ability. Such systems track students' performance and provide feedback to them, hence leading students who encounter weakness in a specific subject through conventional instruction methods. There has also been empirical evidence showing that AI-based personalized learning creates a deeper understanding of the subject material as it addresses the variability in students' learning style (Lokare & Jadhav, 2024). In this, AI algorithms can detect whether a student learns by way of visual, auditory, or kinesthetic action and modulate the content of instruction accordingly. This flexibility is relevant in an inclusive classroom because there are varying needs and capacities among different students.

AI driven-assistive tools play a crucial role in facilitating the engagement of learners with disabilities through effective interaction with learning activities (Ngalim, 2019). This is achieved by assisting the accessibility of content for learners and interactions with the classroom environment. Significant benefits for students who have communication or language disabilities could include speech recognition and TTS through AI, for instance, where tools like Google Live Transcribe, as well as Microsoft Immersive Reader, transcribe spoken words into print in real time to benefit hearing-impaired students. Comparable communication assistant tools utilizing AI, including augmentative and alternative communication (AAC), facilitate the inability or reduced ability of a student with a communication disorder to be more visibly interactive in class. AI has been equally instrumental for students with physical disabilities. For instance, eye-tracking software can be used to empower students with restricted mobility in navigating digital platforms and interacting with learning content using only their eyes (Madhusanka et al., 2024). The use of AI-empowered robotic systems enables students with physical impairments to participate in other activities offered in the classroom setting that could otherwise be inaccessible to them. In the case of students with cognitive disabilities, AI provides tools that scaffold learning and allow for understanding. The use of natural language processing in text simplification systems by analyzing documents, making hard-to-read material easy to understand, bridges the gap between the individual capacity of students in reading and what the curriculum demands.

AI builds on the ability of an educator to use data in real time to differentiate instructions for students. Learning analytics in education provides information about the student's progression, strengths, and weaknesses; and so the teacher can intervene appropriately. Predictive analytics with AI power predicts student performance and identifies any potential behind-time failures. AI systems can scan the history data of students to identify patterns that may signal potential academic learning problems and allow for intervention in plenty of time to intervene accordingly. This is very helpful in inclusive classrooms where students with learning difficulties will be identified early before they fall too far behind. AI-powered platforms give teachers fast and accurate feedback about their students' engagement and understanding. For example, facial recognition software and emotion detection systems can read the students' facial expressions and body language to tell if a student is engaged or not in class (Calvo et al., 2020). Such insight helps the teacher change the teaching plans promptly to keep the students engaged and change the curriculum before frustration or confusion set in.

AI can especially remodel how curricula are designed and delivered to make it more representative of the diversity of needs of the students. AI can be used in tailoring content to meet the specific learning preferences and abilities of individuals within a classroom, thus ensuring that the learning environment is inclusive. Inclusive education presupposes differentiation and AI provides the resources needed to differentiate instruction on different levels. Studies show that through AI, it is possible to automate the differentiation process by changing the complexity of content, switching up learning tasks, and adjusting instructional strategies according to the specific needs of students (Gligorea et al., 2023). This, in return, leaves the teachers with an opportunity to attend to the engagement and specific challenges that arise within the classroom. AI-powered technologies on gamification and immersive learning make new avenues where various students of diverse abilities and learning styles can be engaged (Mahmoudi-Dehaki et al., 2024). It has been proven that game-based learning environments increase motivation and participation among students, more so with regards to those who have learning disabilities or attention difficulties. Likewise, with VR and AR technologies, there can be virtual reality scenarios that depict abstract concepts, including other applets that can offer an immersive experience, satisfying various learning styles.

3.3 Application of AI in Inclusive Education

AI applications in inclusive education include the following: (a) assistive technologies as per the needs of each student, (b) personalized learning, and (c) inclusive assessment tools (IATs) and intelligent tutoring systems (ITSs).

3.3.1 Assistive Technologies as per the Needs of Each Student

AI is essentially a forerunner in rendering inclusive education using assistive technologies to cater to the needs of every child, wherein the needs of children suffering from disabilities or learning disorders are taken care of (Matabane, 2022). In this way, AI solutions remove the barriers from the process of learning by providing access to tailored solutions that suit the physical, cognitive, and sensory requirements of students. Learners who suffer from speech or language disorders require TTS and STT technology that operates on AI. Tools like the Microsoft Immersive Reader convert text into an audio stream and help students who have reading disabilities. For instance, environments like speech recognition support any students who have speech disabilities and can type in assignments or other activities. AI-enabled eye-tracking software and robotic systems support students who have mobility impairment. It is possible to use digital content using one's eyes as students with minimal mobility have utilized eye-tracking technology. AI-based robotic support can assist students in doing tasks that demand physical skills they do not possess. The domain of natural language processing (NLP) is facilitating cognitive support for these students through text simplification systems that adjust the readability level, thereby making the reads easier for intellectually disabled children (Barbu et al., 2015). Other support will be found in the form of cognitive scaffolding tools where some rough concepts are divided into steps for comprehension. These artificial devices, then, powered by AI, allow students with communication disorders to communicate through symbols, texts, or synthesized speech, thereby conveying their needs without the use of spoken language. And since these tools allow for interaction in the classroom, they allow for self-expression as it follows from participation with peers and teachers alike.

3.3.2 Personalized Learning

One of the areas that artificial intelligence is currently transforming in the field of education is inclusive education, customizing learning experiences for diverse students irrespective of their abilities, backgrounds, or learning preferences (Mohammed & Nell'Watson, 2019). AI technology tools and platforms apply data-driven insights to individualize instruction with real-time adjustment of content and learning paths. These technologies enable educators to better meet the needs of learners in inclusive classrooms, where learning abilities are diverse. Among the most widely used applications through which AI manifests in personalized learning are adaptive learning systems. The platforms, namely DreamBox and Smart Sparrow, analyze student-performance data, adjust the difficulty of tasks, and pacing the methods of instruction based on each learner's performance (Ezzaim et al., 2022). This challenges students appropriately, neither leaving some behind nor disengaging them from the material. For children with learning disabilities or other challenges, this tailored approach can significantly enhance engagement and comprehension. AI also promotes assistive technologies that facilitate personalized learning opportunities for students with disabilities. For instance, support for TTS and STT, would allow a student with difficulties in reading or writing (Nordström et al., 2019). AI-based communication aids such as those provided by augmentative and alternative communication would help students who are non-verbal to express themselves. Similar to pupils with physical disabilities, eye-tracking technology, for instance, enables them to interact independently with digital content. Additionally, learning analytics based on artificial intelligence provide teachers with immediate feedback on students' learning actions so that teachers can intervene with pupils at the point when they are experiencing difficulties and harmonize teaching strategies to overcome their weaknesses. Such an approach provides more effective teaching in an inclusive classroom because it is able to identify the specific learning gaps of individuals or groups of learners and adjust the curricula in these directions.

3.3.3 Inclusive Assessment Tools (IATs) and Intelligent Tutoring Systems (ITSs)

Inclusive assessment tools employ artificial intelligence for alternate and accessible means of assessing students' learning progress. On the other hand, IATs are comparable alternatives that, to date, allow the accurate assessment of a child with a disability or learning difference (Buttrick et al., 2020). Examples include the STT and TTS tools that aid children with dyslexia; voice recognition is applicable when assessing understanding by children with a physical impairment who cannot write. AI-based systems provide unbroken, real-time feedback to teachers regarding their approaches while teaching students based on the performance of the individual student. AI also supports formative testing, wherein, instead of making a one-time test, constant assessment is provided to the students, which tends to remove stress and increase the efficiency of learning (Oladele & Ndlovu, 2021). Intelligent tutoring systems, through AI, provide instruction tailored according to the requirements of the student. The system can even identify the learning style and pace of a student as well as knowledge gaps, and adapt accordingly. ITS systems, such as Carnegie Learning's "Cognitive Tutor," allow for differentiated instruction; students who are advanced will receive complex content that pushes their reasoning and solution skills, while students who have a learning need can receive additional support. ITS also affords immediate opportunities for feedback so that the student can instantly correct errors and strengthen concepts most effectively. Further capabilities enable education access for disabled students or those with a learning difference, permitting them to learn independently.

3.4 Benefits of AI in Inclusive Education

The benefits of AI in inclusive education include the following: (a) enhanced accessibility, (b) data-driven decision-making, (b) improved engagement, (d) early intervention, and (e) continuous enhancement.

3.4.1 Enhanced Accessibility

AI greatly facilitates accessibility in education through innovative tools breaking down barriers to students with disabilities or learning differences. The ability of AI to provide individualized support means all students with varying abilities can fully participate in the process and have an equal chance of success (Jaiswal & Arun, 2021). AI-based assistive technologies offer additional means of input that benefit students who have physical impairments. For example, eye-tracking software and voice-command systems allow students who cannot make use of computers to freely interact with learning resources. AI-driven robots enable more straightforward participation by such students in activities that would not have been possible without these robots (Srinivasa et al., 2022). Such technologies empower students with disabilities to become increasingly independent, giving them equal access to education. AI helps provide for the inclusion of students with speech and language impairments. Speech-to-text (STT) and text-to-speech (TTS) technologies, for instance, Google Live Transcribe and Microsoft Immersive Reader, have allowed children with a hearing or speech impediment to be able to converse and read by having content presented in writing (Yeh, 2014). Language translation tools, backed by AI, tend to make learning easier for non-English speakers to learn in their native language simultaneously, gradually increasing fluency in the language to be learnt, cutting linguistic barriers. AI has made learning accessible to everybody as it creates customized materials for various kinds of cognitive abilities. For example, AI can make complicated texts easier to understand, or even add pictures to facilitate the process of comprehension about abstract concepts for students with intellectual disabilities. The adaptive learning platforms automatically adapt the content that's being presented to a student based on that student's performance, so they learn at their own pace and in their preferred style.

3.4.2 Data-Driven Decision-Making

AI plays an important role in enhancing inclusive education as data-driven decision-making empowers educators with more information that enables them to tailor teaching strategies toward the diverse needs of learners (Shivashankar & Bakthavatchaalam, 2024). AI-based systems collect data in real time regarding student performance, behaviors, and learning patterns. Such constant feedback enables educators to keep track of learners' improvements, especially in the case of those who require special attention due to disabilities or learning differences. Teachers can intervene early by making changes to instruction strategies and targeted support when their students show weaknesses in certain areas or strengths in others. AI will look at historical and current student data to determine trends and outcomes for future academics. For example, through machine learning algorithms, one can detect early signs of academic disengagement or learning difficulties, which will then allow educators to take proactive measures against problems before they worsen. This is very practical in inclusive education as, at the point of falling behind their peers, a student with learning disabilities should already have received early intervention. Using data analytics, AI supports the creation of personalized learning experiences tailored to each pupil's needs. The analysis of pupil-level data—including how quickly a pupil learns, how well he or she understands the content, and his or her level of engagement—conducted on AI-driven platforms determines whether the customized educational materials are apt for the individual pupil (Aicardi et al., 2018). For those students with disabilities, it is essential for differentiated instruction whereby access is balanced across all learning. AI tools give teachers an insight into where to further refine their curricula and how to make adjustments for the inclusion of more diverse learners; devised lesson plans and teaching will therefore be more inclusive, considering different learning styles, abilities, and backgrounds.

3.4.3 Improved Engagement

The biggest way in which AI increases engagement in the classroom is through personalized learning. AI-based learning platforms can customize learning content and pedagogy to suit the abilities, skills, and pace of individual learners (Kaswan et al., 2024). Such methods ensure that students stay engaged because they provide students with challenges on a level suitable for them at a particular time. Advanced learners can be challenged through more complex topics, while more support is given to those who need it, hence not leaving any learner behind or under-stimulated. More importantly, AI enables gamification and interactive learning to be added, which is most effective in keeping the interest of students. Games, simulations, and quizzes powered by AI are all interesting ways to deliver learning, hence producing a positive emotional connection with the material (Westera et al., 2010). This is particularly helpful for students who have learning disabilities, as they will not be as challenged as when using traditional approaches. AI reacts immediately, thus helping students to realize their errors and correct them in real time. It is through this responsiveness that the learner stays attentive since their correct answers are enhanced while they are being gradually led through any challenges. For the disabled and those with learning difficulties,

AI contains speech recognition and text-to-speech capabilities that enable them to participate more fully. Thus, it enhances the sense of belonging and inclusion.

3.4.4 Early Intervention

The role of AI in inclusive education is definitely transformational, because it actually identifies learning difficulties and developmental delays early on, thus ensuring timely action and preventing educational disadvantages in the long run. Early intervention is also important because it means the earlier a learning or developmental problem is detected, the more support and intervention can be offered (Scott & Baldwin, 2004). The analysis of enormous amounts of data—student behavior, performance patterns, and engagement levels—can also be done using AI in order to identify early warning signals for learning problems or developmental difficulties. Using AI-driven tools such as learning analytics systems, the learning progress of students can be monitored in real time. Moreover, the anomalies that may result in challenges, for instance, may be highlighted. For example, when a student cannot read with comprehension or is experiencing problems with understanding instructions, AI systems can raise red flags about those problems. Such early detection allows instructors to quickly intervene and apply focused intervention before the student lags behind. AI also assures early interventions by monitoring the effects of interventions over time. As AI produces real-time information on a student's progress, it helps educators make adjustments and informed choices based on data that will eventually translate to better, dynamic support for students who are better prepared for long-term results (Owan et al., 2023).

3.4.5 Continuous Enhancement

One very important role AI will play to come up with more innovative forms of inclusive education is the continuous provision of support and improvement that will particularly be aligned to the needs of individual learners (Xia et al., 2022). Dynamic adaptation regarding students, educators, and institutions benefits students, educators, and educational institutions in many fundamentally different ways. AI systems are designed to continually analyze students' performance data that include strengths, weaknesses, and learning preferences. This allows educators to revise their strategies and resources in real time as each student receives support. For instance, advanced AI can dynamically change the learning pathway as students advance. It will provide more challenging material or other resources based on their needs. The immediate feedback tool in AI is a great help to students to understand what they have done wrong and how they can improve it. The immediate response cultivates a growth mindset, motivating the student to self-direct their learning process by taking total responsibility for it. Educator real-time analytics can deepen the understanding of class dynamics. It can afford educators timely interventions for struggling students; AI systems can directly interact with a huge number of students, hence making it easier for educators when handling diverse classrooms (Chen et al., 2023). Scalability ensures that one-on-one attention is also still possible in inclusive settings where resources are limited. AI may automate a great deal of administrative work, giving teachers more time to spend on teaching and less on paperwork. AI harvests and processes a huge amount of data regarding student performance, engagement, and learning style. Such an approach rooted in data allows teachers and administrators the opportunity to take well-informed decisions on curricula, resources provision, and professional development (Means et al., 2009). On top of knowing which strategies will suit which types of learners, educational institutions can continue to hone their practices.

3.5 Challenges and Ethical Issues of AI in Inclusive Education

The integration of AI in inclusive education offers significant opportunities for personalized learning and improved outcomes. However, it also raises several challenges and ethical issues that must be addressed to ensure equitable and effective implementation. Below, the primary challenges related to data privacy, data security, teacher training and support, costs, and access are discussed.

3.5.1 Data Privacy

AI systems require large quantities of data to operate and function well, which include sensitive student information, such as learning disabilities, behavioral patterns, and personal identifiers (Akgun & Greenhow, 2022). This raises critical issues related to data privacy, because improper or unauthorized use or access can lead to serious violations of confidentiality. Strict regulations apply with respect to student data as dictated by legislation including the Family Educational Rights and Privacy Act in the US. There should be rules in schools that apply to those laws as long as AI systems are involved. Well-defined policies on how the data are collected and circulated should be implemented. Furthermore, informed consent from parents and guardians is also a requirement but could be an inconvenience, especially in instances where there are more young students or students with disabilities.

3.5.2 Data Security

As we become increasingly dependent on digital equipment, cyberattacks and data breaches are also expected to increase. Thus, schools and institutions associated with education need to have the most advanced security measures to protect sensitive student-related information from being hacked into and exploited (Liesen, 2017). AI systems, if not secured properly, become an attractive target for hackers as well, through which personal data can also be compromised; much more seriously, the integrity of the sources for learning can be questioned. Data security requires constant investment in technology and training, thereby adding more of a burden to educational institutions.

3.5.3 Teacher Training and Support

AI for inclusive education means teachers need to be trained enough in using these technologies in classrooms. In that sense, there is an obvious limitation: much of this program would not be able to be used effectively by many teachers because of a lack of technical skills and/or knowledge in the uses of all this software. Teachers should be offered professional development programs showing them the best way of handling AI technologies, analyzing the data analytics, and leveraging adaptive learning tools (Colchester et al., 2017). In the case that they are not supported, there is a high probability that AI technologies will be underutilized, thus limiting their maximum advantages for the diverse learners.

3.5.4 Costs

AI technology in education can be very expensive to implement since the creation, management, and upgrading of AI technologies require significant funding. Limited budget schools, particularly those in under-funded school districts, may be unable to afford to set aside money for AI tools, training, and supporting funds. The cost of purchasing or licensing AI systems means institutions may be less likely to take action and invest in these technologies (Teece, 2018). This financial limitation also perpetuates the existing inequalities by creating an odd circumstance that holds out ripe opportunities for some schools to be disadvantaged because of their low resources.

3.5.5 Access

The digital division remains the most significant education-related problem. Not all students benefit from technology equity, including devices and high-speed internet access, without which AI tools cannot operate (Muwani et al., 2022). This access disparity in itself may marginalize already vulnerable students from low-income families or rural areas, thereby intensifying the achievement gap. Thus, AI technologies must be ensured to equalize access so that education may become more inclusive. For inclusive education schools have to scuffle to provide the necessary infrastructure and devices, along with connectivity to the internet, as a prerequisite for students to understand and benefit from AI in the learning process.

3.6 Conclusion

The assimilation of AI into inclusive education has enhanced learning by developing engagement, personalized instruction, and accessibility. AI-directed technologies, adaptive learning systems, and advanced learning equipment generate opportunities for all types of students including those with differently abled abilities to get tailored learning experiences. The advantages of AI in education go beyond individual students, giving educators data-based knowledge to enhance teaching strategies and develop a sense of overall educational output. However, challenges like data privacy issues, teacher training, high costs, and equitable access need to be resolved to ensure the ethical and efficient utilization of AI in education. While resolving all these issues in AI-driven invention, the education sector needs to ensure that all students, irrespective of their abilities or backgrounds, have an equal chance to thrive in the digital age.

3.7 Future Direction

The integration of AI in inclusive education should consider not only current needs but also future developments for educational practices. This study emphasizes key areas such as R&D in AI for inclusive education, policy and regulatory frameworks, and necessary collaboration and partnerships among many stakeholders. These aspects must be better understood by educators, practitioners, and policymakers to deploy AI and new technologies effectively. Below more detailed descriptions of these future directions are provided.

3.7.1 AI R&D for Inclusive Education

Investment in AI R&D is required to further advance the application of AI in support of inclusive education (Pham & Sampson, 2022). Future research areas include the development of AI tools specifically designed with diversity in learning needs, such as students with disabilities or with differences in cognitive ability. In this sense, adaptive learning platforms that can adapt in real time to the student's interactions with content based on their responses need to be developed. Longitudinal studies must be conducted regarding the effects on students' outcomes, engagement, and the overall experience. It is by knowing what works and in what circumstances that AI tools will be fine-tuned to better meet classroom requirements. Researching the ethical frameworks that guide the use of AI in education, foregrounding privacy, equity, and transparency, will ensure that applications for AI will not present or perpetuate bias and respect the dignity of students.

3.7.2 Policy and Regulations

It is thus adequate policies and regulations that signify whether AI is used appropriately for effective and ethical learning in inclusive education. This will also imply some of the considerations, such as developing comprehensive regulations that guide the collection, storage, and sharing of data practices in educational settings; policies should also go as far as establishing methods for maintaining student information private with a clear protocol on data access; developing standards and guidelines for AI technology in evaluating the effectiveness, accessibility, and inclusivity of education; work with experts in education to define benchmarking that can be used to monitor AI tools; advocate for government and institutional funding to purchase and implement AI technologies, especially in underfunded schools.

3.7.3 Cooperation and Coordination

Competition among the various stakeholders will feature significantly during the AI implementation process into inclusive education. It encourages collaboration among education institutions, technology developers, researchers, and non-profit organizations to better develop innovation and best practices. A collection of these collaborations could co-create different tools that would help solve quite a few problems in education. They should implement collaborative training programs for educators, which will better prepare them to use AI technologies in their teaching environments. Training programs should be constantly developed for teachers in order to equip them with needed expertise and materials. The application of AI in education should also be publicized to the community and families. Stakeholders will be able to contribute to the development of even more practical and efficient tools that cater to the different needs of students.

3.8 Implications for Educators, Practitioners, and Policymakers

The findings and implications of this study have several very important consequences. These include the findings from this study informing educators and practitioners on smart decision-making concerning the adoption and implementation of AI tools, helping them to better understand the strengths and weaknesses of AI in all respects that would guide them to the correct technologies that are potentially aligned with the learning goals. Policymakers may formulate strategies toward equity of access and outcomes in education so that every student will be able to enjoy a high-quality experience irrespective of their cognitive, behavioral, or sociocultural characteristics, when facilitated by AI. By providing R&D and collaboration, stakeholders can propel innovation in the education practices and ensure that students' living experience is dynamic, along with adjustment to continuously changing needs in the most effective and appropriate manner.

References

Aicardi, C., Reinsborough, M., & Rose, N. (2018). The integrated ethics and society programme of the Human Brain Project: Reflecting on an ongoing experience. *Journal of Responsible Innovation*, *5*(1), 13–37.

Ainscow, M. (2020). Promoting inclusion and equity in education: Lessons from international experiences. *Nordic Journal of Studies in Educational Policy*, *6*(1), 7–16.

Almgren Bäck, G., Lindeblad, E., Elmqvist, C., & Svensson, I. (2024). Dyslexic students' experiences in using assistive technology to support written language skills: A five-year follow-up. *Disability and Rehabilitation: Assistive Technology*, *19*(4), 1217–1227.

Alharbi, W. (2023). AI in the foreign language classroom: A pedagogical overview of automated writing assistance tools. *Education Research International*, *2023*(1), 4253331.

Akgun, S., & Greenhow, C. (2022). Artificial intelligence in education: Addressing ethical challenges in K-12 settings. *AI and Ethics*, *2*(3), 431–440.

Bambrick-Santoyo, P. (2010). *Driven by data: A practical guide to improve instruction*. John Wiley.

Barbu, E., Martín-Valdivia, M. T., Martinez-Camara, E., & Urena-López, L. A. (2015). Language technologies applied to document simplification for helping autistic people. *Expert Systems with Applications*, *42*(12), 5076–5086.

Buttrick, N., Axt, J., Ebersole, C. R., & Huband, J. (2020). Re-assessing the incremental predictive validity of Implicit Association Tests. *Journal of Experimental Social Psychology*, *88*, 103941.

Calvo, R. A., & D'Mello, S. (2020). Affect detection: An interdisciplinary review of challenges and opportunities. *IEEE Transactions on Affective Computing*, *11*(1), 4–24.

Chen, Y., Jensen, S., Albert, L. J., Gupta, S., & Lee, T. (2023). Artificial intelligence (AI) student assistants in the classroom: Designing chatbots to support student success. *Information Systems Frontiers*, *25*(1), 161–182.

Colchester, K., Hagras, H., Alghazzawi, D., & Aldabbagh, G. (2017). A survey of artificial intelligence techniques employed for adaptive educational systems within e-learning platforms. *Journal of Artificial Intelligence and Soft Computing Research, 7*(1), 47–64.

Ezzaim, A., Kharroubi, F., Dahbi, A., Aqqal, A., & Haidine, A. (2022). Artificial intelligence in education-State of the art. *International Journal of Computer Engineering and Data Science (IJCEDS), 2*(2), 1–11.

Gligorea, I., Cioca, M., Oancea, R., Gorski, A. T., Gorski, H., & Tudorache, P. (2023). Adaptive learning using artificial intelligence in e-learning: A literature review. *Education Sciences, 13*(12), 1216.

Griffiths, T., Slaughter, R., & Waller, A. (2024). Use of artificial intelligence (AI) in augmentative and alternative communication (AAC): Community consultation on risks, benefits and the need for a code of practice. *Journal of Enabling Technologies, 18*(4), 232–247.

Halkiopoulos, C., & Gkintoni, E. (2024). Leveraging AI in E-learning: Personalized learning and adaptive assessment through cognitive neuropsychology—A systematic analysis. *Electronics, 13*(18), 3762.

Jafarnia, A., Hariri, H., & Parvizi, G. R. (2023). Unlocking the potential: Exploring the multifaceted impact of artificial intelligence integration in language learning. *Language Education and Technology, 3*(2), 173–185.

Jaiswal, A., & Arun, C. J. (2021). Potential of Artificial Intelligence for transformation of the education system in India. *International Journal of Education and Development using Information and Communication Technology, 17*(1), 142–158.

Kaswan, K. S., Dhatterwal, J. S., & Ojha, R. P. (2024). AI in personalized learning. In Adarsh Garg, B V Babu, Valentina E Balas (Eds.) *Advances in Technological Innovations in Higher Education* (pp. 103–117). CRC Press.

Liesen, J. J. (2017). *Qualitative case study exploring operational barriers impeding small and private, nonprofit higher education institutions from implementing information security controls* [Doctoral dissertation]. Northcentral University.

Lokare, V. T., & Jadhav, P. M. (2024). An AI-based learning style prediction model for personalized and effective learning. *Thinking Skills and Creativity, 51*, 101421.

Madhusanka, B. A., Ramadass, S., Rajagopal, P., & Herath, H. M. K. K. M. B. (2024). *Artificial Intelligence-Based System for Gaze-Based Communication*. CRC Press.

Mahmoudi-Dehaki, M., & Nasr-Esfahani, N. (2024). Harnessing the power of persuasion in education: Persuasive gamification and AI. In Mageswaran Sanmugam, Darren Lim, Nurul Maziah Mohd Barkhaya, Wan Ahmad Jaafar Wan Yahaya and Zuheir Khlaif (Eds.) *Power of Persuasive Educational Technologies in Enhancing Learning* (pp. 177–204). IGI Global.

Matabane, M. B. (2022). *Crisis in the implementation of inclusive education: Reflection on the role of full-service schools in Sekhukhune district* [Doctoral dissertation], University of Limpopo.

Means, B., Padilla, C., DeBarger, A., & Bakia, M. (2009). Implementing data-informed Qdecision making in schools: Teacher access, supports and use. *US Department of Education, 1*, 1–110.

Mohammed, P. S., & 'Nell'Watson, E. (2019). Towards inclusive education in the age of artificial intelligence: Perspectives, challenges, and opportunities. *Artificial*

Intelligence and Inclusive Education: Speculative Futures and Emerging Practices, 1, 17–37.

Muwani, T. S., Ranganai, N., Zivanai, L., & Munyoro, B. (2022). The global digital divide and digital transformation: The benefits and drawbacks of living in a digital society. In Munyaradzi Zhou, Gilbert Mahlangu, and Cyncia Matsika (Eds.) *Digital Transformation for Promoting Inclusiveness in Marginalized Communities* (pp. 217–236). IGI Global.

Nordström, T., Nilsson, S., Gustafson, S., & Svensson, I. (2019). Assistive technology applications for students with reading difficulties: Special education teachers' experiences and perceptions. *Disability and Rehabilitation: Assistive Technology, 14*(8), 798–808.

Ngalim, V. B. (2019). Lack of human resources as barriers to the implementation of inclusive education in the University of Bamenda, Cameroon. *European Journal of Special Education Research, 4*(3), 41–58.

Oladele, J. I., & Ndlovu, M. (2021). A review of standardised assessment development procedure and algorithms for computer adaptive testing: Applications and relevance for Fourth Industrial Revolution. *International Journal of Learning, Teaching and Educational Research, 20*(5), 1–17.

Owan, V. J., Abang, K. B., Idika, D. O., Etta, E. O., & Bassey, B. A. (2023). Exploring the potential of artificial intelligence tools in educational measurement and assessment. *Eurasia Journal of Mathematics, Science and Technology Education, 19*(8), em2307.

Patel, T., Dorff, J., & Baker, A. (2022). Development of special needs classroom prototypes to respond to the sensory needs of students with exceptionalities. *Archnet-IJAR: International Journal of Architectural Research, 16*(2), 339–358.

Pham, S. T., & Sampson, P. M. (2022). The development of artificial intelligence in education: A review in context. *Journal of Computer Assisted Learning, 38*(5), 1408–1421.

Scott, J., & Baldwin, W. L. (2004). The challenge of early intensive intervention. In Angi Stone-MacDonald, David F. Cihak, Dianne Zager, Dianne Zager (Eds.) *Autism Spectrum Disorders* (pp. 183–238). Routledge.

Shivashankar, K., & Bakthavatchaalam, V. (2024). Education policies through data driven decision making: Accelerating inclusive education for people with disabilities. *Artificial Intelligence Enabled Management: An Emerging Economy Perspective, 1*, 15–32.

Srinivasa, K. G., Kurni, M., & Saritha, K. (2022). Harnessing the power of AI to education. In K. G. Srinivasa, Muralidhar Kurni, Kuppala Saritha (Eds.) *Learning, Teaching, and Assessment Methods for Contemporary Learners: Pedagogy for the Digital Generation* (pp. 311–342). Springer Nature Singapore.

Teece, D. J. (2018). Profiting from innovation in the digital economy: Enabling technologies, standards, and licensing models in the wireless world. *Research Policy, 47*(8), 1367–1387.

Xia, Q., Chiu, T. K., Lee, M., Sanusi, I. T., Dai, Y., & Chai, C. S. (2022). A self-determination theory (SDT) design approach for inclusive and diverse artificial intelligence (AI) education. *Computers & Education, 189*, 104582.

Westera, W., Prada, R., Mascarenhas, S., Santos, P. A., Dias, J., Guimarães, M., ... & Ruseti, S. (2020). Artificial intelligence moving serious gaming: Presenting reusable game AI components. *Education and Information Technologies, 25*, 351–380.

Woolf, B. P. (2010). *Building Intelligent Interactive Tutors: Student-Centered Strategies for Revolutionizing e-Learning*. Morgan Kaufmann.

Yeh, R. (2014). *Effective Strategies for using Text-to-Speech, Speech-to-Text, and Machine-Translation Technology for Teaching Chinese: A Multiple-Case Study*. Northcentral University.

4

Artificial Intelligence Tools for Instructors and Learners to Optimize the Teaching and Learning Processes

Pallavi Upadhyay, Harshit Bhardwaj and Pradeep Tomar

4.1 Introduction

The development of AI brings a transformation to industries, businesses, commerce, and education, where education is one of the most affected sectors. AI applications in education are transforming the organizational, secretarial, and administrative tasks that concern teaching and the creation of adaptive learning environments. By automating repetitive tasks, providing real-time data visions, and enhancing student interactions, AI frees up educators to concentrate more on mentorship and direct instruction. The AI tools available to students make learning diversified and flexible, responding to the variety of different learning needs and styles that result in better understanding and retention.

Academically, AI is now a pivotal resource that provides a wide array of sophisticated tools and methodologies to be utilized in terms of improving teaching as well as learning. AI-driven platforms provide educators with streamlined work in grading and other paperwork, allowing them to spend more quality time with students. Personalized learning programs are developed by adaptive learning platforms to guarantee that each student progresses according to a timetable that best suits their needs and understanding. Intelligent Tutoring Systems (ITSs) are examples of in-person tutoring sessions that provide learners with immediate feedback and assistance when they face actual learning obstacles. Learning analytics reinforce the types of skills AI might provide by giving students insightful feedback on their performance and engagement through data interpretation, enabling educators to identify struggling learners before it becomes a serious problem [1].

Natural language processing-based artificial intelligence-powered grading and feedback systems increase consistent grading and instantaneous

DOI: 10.1201/9781003560814-4

assessment of student writing, increasing evaluation efficiency and uniformity. Students' engagement has been effectively raised by gamification tactics that include game aspects in educational procedures, making learning more engaging and enjoyable. Additionally, as NLP technology has advanced, language learning may now include instant feedback on grammar and pronunciation, greatly enhancing language acquisition through interactive practice. AI-assisted content creation and recommendation tools enable educators to better tailor lesson plans and resources to students' varied learning needs, increasing the curriculum's effectiveness and adaptability [2].

With all of these developments, AI improves learning experiences, increases student engagement, relieves educators of some of their workload, and produces better educational outcomes. By doing this, technology creates a dynamic and adaptable learning environment that benefits students and gives educators more authority [3].

4.2 AI Tools for Teaching and Administrative Support

Educators may create instructional materials more quickly and efficiently with the help of AI-powered content creation tools. Lesson plans, tests, and assignments may be created using tools like ChatGPT and Jasper according to predetermined criteria. For example, when educators enter a topic such as "Introduction to Machine Learning," the AI will provide material summaries, important definitions, and even practice problems that are appropriate for the level of difficulty of the curriculum. This speeds up the development of the course and makes it possible to quickly modify the course contents to accommodate changing student demands. In the past, the main goal of these AI-powered teaching tools was to improve the utilization of resources, assessment, grading, and material so that educators could devote more time to pedagogy and student engagement.

- Intelligent Tutoring Systems (ITSs): These use AI to give students feedback and instructional help that is tailored to their individual learning progress. ITSs are able to evaluate the student's response and identify any areas in which the learner can benefit from more practice or training. ITSs are made to offer one-on-one coaching by adapting the course content to the student's level of understanding and pace. For instance, Carnegie Learning and Smart Sparrow's solutions enable individualized feedback and learning pathways with real-time student data analysis, allowing the content to take on several forms. As a result, we may view ITSs as crucial instruments that help educators meet the various requirements of their pupils without the need for human interaction. This is the reason why ITSs look forward to a more personalized learning trajectory that would increase engagement as well as improve the results of learning. Such systems may also prove to be almost one-to-one coaching in enabling enhanced explanation, clarification, and adaptation, thus allowing for a more customized and dynamic learning environment. Some of the varied courses using ITSs are physics and mathematics in which adaptive help makes complicated concepts easier to be understood [4–6].

- Learning Analytics: AI is used in learning analytics where student performance data are analyzed in a manner that informs educators of what is happening in the classroom. Analytics can track scholars or areas of problems for students and supports them with early intervention and development. For instance, with Microsoft's Tableau and Azure Analytics, student engagement, involvement, and academic performance can be tracked, making it possible to design interventions and courses specifically tailored to each student. Learning analytics makes use of the data that students create during their education to gain an insight into the process. These AI-based learning analytics monitor how students interact with materials and identify trends that point to both areas of strength and trouble. They allow educators to make fact-based decisions about how to structure classes to either reinforce scholar's or student's strengths or overcome their weaknesses. Learning analytics may combine data from various sources to provide trends, which indicate larger patterns, in addition to tracking individual development. Such uses might assist educational establishments in refining their curricula and creating effective teaching methods [5,7].
- Automated Grading and Feedback: Gradescope's AI-powered technology provides constant feedback and automatically assesses responses to objective tasks. As a result, teachers spend more time evaluating higher-order thinking skills and engaging learning activities while cutting down on assessment time. An automated assessment system often uses machine learning and NLP to evaluate students' responses, especially when they are completing predefined forms such as short essays or objective response activities. Students receive immediate feedback from this method, which is crucial for learning since it makes it simpler for them to recognize and fix errors as soon as they occur. Teachers will have more time to design additional classes, spend much-needed time with pupils, and do less repetitive work thanks to automated grading. For instance, some evaluation tools evaluate written replies and offer feedback on the response's consistency, grammar, and quality; studies have shown that these features improve students' performance over time [8].
- Virtual Teaching Assistants (VTAs): In brief, VTAs assist students with debates, problem solving, and concept clarification. AI has made it possible

for these assistants to be available around-the-clock and offer consistent assistance regardless of time or place. They may be contacted by voice, text, or even photos, making learning far more accessible and adaptable. Big language models and machine learning are typically used by VTAs to provide explanations for a variety of enquiries that are appropriate for each user's comprehension level. Because AI handles routine queries and teachers may modify their lessons to meet the requirements of various pupils, this greatly reduces the strain on educators. These digital assistants, such as Google Dialogflow and IBM Watson Education, may assist students with their enquiries around-the-clock, carry out administrative duties, and respond to common queries. Because this gateway provides immediate feedback, it makes the classroom more accessible and encourages participation, fostering an even more encouraging learning atmosphere [9].

4.3 AI Tools for Personalized Learning

With the aid of AI, learning could be provided to a person who wishes it to be tailored to a desired learning style, and students can then study at their own speed in their own way. One of the most important strategies in meeting the various demands of students is the use of personal learning resources.

- Adaptive Learning Platforms: These are AI-driven systems which provide a customized experience of learning based on varied learning paths, pace and content, focused on catering to the diverse needs of each learner and carrying out an analytical evaluation of their progress. Using machine learning algorithms, these systems are able to gauge the level of performance achieved in the present. This can then be used to adjust the difficulty of the content, but also to steer students to needed resources in particular areas, thus building engagement and retention. Many studies have shown that these AI-based adaptive systems in e-learning excel over traditional models when combined with more flexibility and additional support for meeting the specific needs of unique individuals. Adaptive systems like Knewton or ALEKS are also used to supplement tougher topics as needed and supplement instruction in STEM subjects. By using adaptive learning systems such as DreamBox and ALEKS, which adjust the material based upon the reaction of the individual student, students will have the opportunity to master easy skills but then navigate challenging ones. This way, the approach enhances learning and ensures student's engagement with learning, with help provided for each at their level of knowledge [10].
- Content Creation/Planning/Recommendation Systems: AI-based content creation tools have proven to be great tools for instructors looking

to quickly apply information. For instance, according to specifications, ChatGPT and Jasper may generate lesson plans, tests, and assignments. A teacher may enter a topic as basic as "Introduction to Machine Learning," for example, and the AI will generate material summaries, precise definitions, and practice problems according to the level of difficulty of the curriculum. Additionally, this expedites the creation of courses and facilitates the quick modification of course content to meet the evolving requirements of students. Instructors can use AI technology for low-stakes quizzes to assess students' knowledge on a regular basis (formative assessment). Additionally, low-stakes tests have yielded a wealth of data on pupils' development, which helps identify areas in which they need to improve. Because Gradescope, another AI tool, recognizes trends in student responses and provides tailored feedback, it also simplifies grading while freeing up teachers' time. Instead of grading administrative activities, this method enables teachers to concentrate on how to improve their lesson based on data [9,10].

Depending on the student's interests, prior performance, and level of involvement, Coursera and Khan Academy's AI recommendation engines provide relevant materials like videos, articles, or exercises. This increases motivation and a chance for deeper learning since its resource-based delivery appeals to individual students [11].

- NLP for Language Learning: Engagement with NLP technologies has transformed the 'nature of teaching.' It offers interactive usage and gives student feedback on vocabulary, pronunciation, and parsing. Duolingo uses NLP to adjust the learning path for each user, based on their errors and progress. Also, NLP systems make people experience languages through role-playing interactions in another context. This way, it trains the lips and ears to create the needed sentences. Perhaps the potential of NLP may also be reflected in automated testing programs that assess written language competency and give feedback with suggestions on more effective usage of the languages. All these are geared toward making the acquiring of a language accessible and interactive. Tools like Duolingo and Grammarly offer constant evaluation of students' actual usage of the language, thus enabling them to gauge their improvement in vocabulary usage, grammar usage, and even pronunciation. NLP systems assist language learners by determining the common linguistic errors and providing remedies [12].
- Gamification and Engagement: By making the way the instructional material is presented more dynamic and gamified, AI-based gamification increases student engagement. Applications like Classcraft and Kahoot, for instance, employ AI to dynamically balance reward and challenge so that students are motivated and engaged by information that is suitable for their current skill level. Gamification is the process of using game aspects outside of games to increase user

or customer engagement. In the educational setting, gamification may make learning enjoyable and serve as a catalyst for level advancement and reaching "ground-level" learning objectives. It has been demonstrated that gamified platforms with leaderboards, badges, and immediate feedback enhance students' motivation and retention in ongoing education. Platforms like Kahoot and Classcraft, for instance, are organized yet adaptable, and offer rewards to students who participate and achieve [13].

4.4 Predictive Analytics for Student Success

Predictive analytics enables educators to anticipate academic outcomes, allowing for proactive support and intervention.

- Student Performance Monitoring: AI-based programs like BrightBytes and Altitude Learning examine student performance, attendance, and engagement data to produce insights that help teachers spot pupils who may be at danger of falling behind in their learning. This realization aids teachers in creating assistance plans that increase student's chances of success [14].
- Identification of at-risk students: Predictive analytics from Starfish Retention Solutions helps identify kids who are at risk of experiencing more academic difficulties than others. This takes place in real time, and instructors may enable customized initiatives that meet each student's needs by using data-driven interventions [15].
- Behavioral Analytics: In order to forecast student engagement and performance, AI-based behavioral analytics systems monitor their interactions and involvement. Teachers can modify their methods to create a more encouraging learning environment and promote active engagement by using behavioral data analysis [16].

4.5 Comparative Table of AI Tools in Teaching and Learning

Table 4.1 provides a structured comparison of various AI tools used in education, categorized based on their functionality. It highlights key tools for intelligent tutoring, adaptive learning, content creation, analytics, automated grading, virtual assistants, learning management, predictive insights,

TABLE 4.1

AI Tools for Teaching and Learning

AI Tool	Purpose	Features	Example Usage
DreamBox	Adaptive learning	Personalized learning pathways, real-time feedback	Math education for K-8 students
Knewton	Adaptive learning	Data-driven content recommendations, student progress tracking	Higher education courses
Carnegie Learning	Intelligent tutoring	AI-driven instruction, step-by-step problem-solving support	Mathematics tutoring
ALEKS	Adaptive learning	Knowledge-space theory, individualized learning plans	College-level STEM subjects
Quizlet	Study assistance	Flashcards, quizzes, gamification	Language learning, exam prep
Socratic by Google	Homework assistance	AI-powered explanations, OCR for text recognition	Assisting students with problem-solving
Duolingo	Language learning	Gamified exercises, AI-driven lesson adaptation	Learning new languages
Babbel	Language learning	Speech recognition, contextual learning	Conversational language acquisition
Grammarly	Writing assistant	AI-driven grammar correction, style improvement	Academic and professional writing
Hemingway Editor	Writing assistant	Readability analysis, simplification suggestions	Improving writing clarity
Canvas	Learning management	Course management, grading automation	Online and blended learning environments
Blackboard	Learning management	Virtual classrooms, analytics	Higher education institutions
EdPuzzle	Interactive learning	Video-based learning, embedded quizzes	K-12 and higher education
Nearpod	Interactive learning	Collaborative lessons, formative assessment	Real-time student engagement
Civitas Learning	Predictive analytics	AI-driven student success predictions	Higher education student retention
Brightspace Insights	Learning analytics	Student engagement tracking, performance monitoring	Personalized teaching interventions
Gradescope	Automated grading	AI-assisted grading, feedback automation	University-level assessments
Turnitin	Plagiarism detection	AI-powered similarity checks, academic integrity	Research paper submission
Zoom with AI Transcription	Virtual classroom	Live transcription, automated captions	Remote learning and accessibility
Google Meet with AI	Virtual classroom	AI-driven noise cancellation, real-time captions	Online teaching and collaboration

gamification, and AI-enhanced virtual classrooms. Each tool's purpose and key features are summarized to showcase how AI optimizes teaching and learning. This comparative analysis helps educators and learners choose the right AI solutions for personalized and efficient education.

4.6 Ethics and Challenges in AI for Education

Bringing AI into education brings along with it ethical issues in terms of privacy data, fairness, and teacher–student interaction.

- Data Privacy and Safety: The enormous volume of student data that AI systems rely on creates issues with data security and confidentiality. To address these concerns, educational institutions are often regulated by frameworks such as the General Data Protection Regulation (GDPR) and the Family Educational Rights and Privacy Act (FERPA) which ensure that data are handled through secure, transparent and accountable practices [17]
- Bias and Fairness in AI Models: When AI systems are trained on non-representative data, biases may be unintentionally introduced. Eliminating these prejudices is crucial to treating all students fairly, regardless of their background. Interaction between educational institutions and AI developers is necessary to evaluate models for fairness and to minimize algorithmic bias [18].
- Teacher–Student Relationship: Even if AI technologies have simplified administration and enabled personalized learning, there is still a risk of being overly dependent on technology. The human aspect of teaching—empathy, mentorship, and moral guidance—cannot be replaced. AI technology should be used by educators to enhance meaningful relationships with their students, not to replace them [19].
- Transparency and Trustfulness: Education professionals and students must comprehend how AI-produced/developed suggestions are made in order to enable them to make responsible and informed use of these technologies in the classroom. Transparency in data utilization and decision-making is the only way to establish trust in AI-based applications [20].

4.7 Future Prospects and Drifts in AI in Education

Emerging drifts in AI suggest that this fast-evolving technology will advance education further.

- Generative AI for Content Generation: Tools such as ChatGPT of OpenAI and Jasper AI can generate educational content, quizzes, and resources that can reduce some of the time spent on preparation by instructors [21].

- Virtual and Augmented Reality in Learning: AI-enhanced virtual and augmented reality applications, for example zSpace, provide immersive learning environments that will engage students in interactive, real-world scenarios, making complex subjects easier to understand [22].
- Policy and Regulatory Frameworks: The US Department of Education and other regulatory bodies will develop policies pertaining to ethical issues emerging from AI in education to ensure awareness and use of AI responsibly and equitably [23].

4.8 Optimization

For instance, instructors can control AI to optimize teaching and learning processes using AI tools for better personalization of instruction, improved assessment, and effective content creation, thereby greatly benefiting instructors as well as learners.

(i) Personalized Learning Paths: AI tools can analyze individual student data, such as learning habits, strengths, and areas for improvement, to create customized learning experiences. They help students learn at their pace and on specific needs. For instance, AI-based platforms can recommend specific resources to students based on their progress in real time, helping them to strengthen weak areas while progressing faster through familiar concepts. This can improve both engagement and academic performance, particularly in large or diverse classrooms where personalized attention is challenging to provide manually [24,25].

(ii) Intelligent Feedback and Assessment: AI-based assessment software offers prompt, detailed, as well as actionable feedback for students to understand their mistakes better. Automatic grading of all assignments such as essays and problem sets saves educators lots of time and also provides students with immediate insights just when they need to retain and internalize new information. Predictive analytics will pinpoint at-risk students to be identified earlier in order to provide more targeted supports such as extra tutoring or resource packets [26,27].

(iii) Content Generation and Adaptation: AI can assist educators in developing adaptive content, for instance, quizzes and simulations, that change in difficulty as students perform. For example, adaptive learning systems could remove repetition of material that advanced students already know but give more basic resources to those who need them. AI also enables the development of multilingual and interactive content that can make heterogeneous classrooms more inclusive and participatory [24,26].

(iv) Enhanced Team Collaboration and Critical Thinking: Virtual assistants or interactive simulations presented through AI systems provide for a more interactive room by encouraging critical thinking and debate. Uses of AI to brainstorm or even in debates with a neutral AI to ensure moderation, collaboration in project work that enables the best of human imagination and artificial intelligence, among others, foster greater understanding and even enhance team learning [27].

(v) Ethical and Practical Issues: Though there exists transformation potential for AI, its challenges lie in data privacy and potential biases. The educator must be honest in revealing the sources of data collected, ascertain that the privacy regulations are upheld, and be cautious of bias arising out of AI output. The training of educators to use AI tools effectively in not allowing technology to replace the pedagogical foundation of teaching is a critical dimension also [28].

Such AI-driven strategies could be deployed thoughtfully within education institutions to offer better learning opportunities, which are both more engaging and malleable. Education, adopting a holistic approach, is responsive to the needs of its students and grows with them to provide optimal teaching and learning opportunities.

4.9 Conclusion

AI tools benefit instructors and learners. They contribute particularly to enhancing personalized learning, improving teaching efficiency, and supporting predictive analytics for student success. Yet, there are questions of ethical difficulties. This should be done to ensure proper use of AI applications in education. By leveraging AI thoughtfully, educators will be best positioned to optimize teaching and learning experiences for a more inclusive, adaptive, and effective educational environment.

References

[1] S. Z. Salas-Pilco, K. Xiao, and X. Hu, "Artificial Intelligence and Learning Analytics in Teacher Education: A Systematic Review," *Education Science*, vol. 13, no. 9, p. 569, 2023.

[2] U.S. Department of Education, "Artificial Intelligence and the Future of Teaching and Learning: Insights and Recommendations," *Office of Educational Technology*, 2023.

[3] M. Cardona, R. J. Rodríguez, K. Ishmael, "AI in Education: Approaches and Strategies for Educators," *Coursera*, 2024.
[4] B. Marr, "How Artificial Intelligence is Revolutionizing the Education Sector," *Forbes*, Oct. 2022. [Online]. Available: www.forbes.com
[5] "Intelligent Tutoring Systems: Enhancing Learning through AI," *The Princeton Review*, 2024. [Online]. Available: www.princetonreview.com/articles/intelligent-tutoring
[6] M. Leon, "Leveraging Generative AI for On-Demand Tutoring as a New Paradigm in Education," *International Journal on Cybernetics & Informatics*, vol. 13, no. 5, pp. 20–21, 2024. [Online]. Available: https://ijcionline.com/articles/on-demand-tutoring
[7] S. Z. Salas-Pilco, K. Xiao, and X. Hu, "Artificial Intelligence and Learning Analytics in Teacher Education: A Systematic Review," *Education Science*, vol. 13, no. 9, pp. 897–899, 2023.
[8] M. Cardona, R. J. Rodríguez, K. Ishmael, U.S. Department of Education, "Artificial Intelligence and the Future of Teaching and Learning," *U.S. Department of Education*, 2023. [Online]. Available: www.ed.gov
[9] W. Holmes and K. Porayska-Pomsta, *The Ethics of Artificial Intelligence in Education*, Routledge, 2022.
[10] D. Smith, "How Artificial Intelligence is Transforming Education," *EdTech Magazine*, Feb. 2023. [Online]. Available: https://edtechmagazine.com
[11] Center for Teaching and Learning, Stanford University, "AI tools in teaching and learning: Technology-enhanced, Policy, Artificial Intelligence (AI)," *Teaching Commons*, Apr. 21, 2023. https://teachingcommons.stanford.edu/news/ai-tools-teaching-and-learning
[12] D. Smith, "How Artificial Intelligence is Transforming Education," *EdTech Magazine*, Feb. 2023. [Online]. Available: https://edtechmagazine.com. Accessed: Nov. 5, 2024.
[13] "Gradescope," *Gradescope by Turnitin*, 2023. [Online]. Available: www.gradescope.com
[14] P. Murray, "Duolingo: Language Learning Made Fun with AI," *TechCrunch*, Apr. 2023. [Online]. Available: https://techcrunch.com
[15] Dror Ben-Naim, "Bringing Learning & Development into the 21st Century", Nov. 14, 2018. Available: https://smartsparrow.com/resources/ebook/improve-learning-development/
[16] OpenAI, *Teaching with AI: A Guide for Educators*, OpenAI, Aug. 31, 2023. [Online]. Available: https://openai.com/index/teaching-with-ai/.
[17] K. Chandra and R. Gupta, "Starfish Retention Solutions: AI in Student Success and Retention," *Journal of Higher Education Analytics*, vol. 15, no. 2, pp. 45–58, 2023.
[18] Brown, "Exploring Behavioral Analytics to Boost Student Engagement," *Educational Technology Review*, vol. 22, no. 1, pp. 16–22, 2022.
[19] J. Locklear, "Data Privacy and AI in Education: Compliance with GDPR and FERPA," *Data Security Journal*, vol. 30, no. 4, pp. 200–210, 2022.
[20] L. Johnson, M. Paul, and E. Gupta, "Addressing Bias in AI for Fairness in Education," *IEEE Transactions on Learning Technologies*, vol. 15, no. 1, pp. 110–120, 2023.

[21] M. Hunt, "Balancing Technology and Human Interaction in Education," *International Journal of Education Technology*, vol. 29, no. 3, pp. 220–230, 2023.
[22] R. E. Sharif and G. T. Mason, "The Role of Transparency in AI for Educational Trust," *Journal of Educational AI Research*, vol. 10, no. 2, pp. 88–96, 2023.
[23] G. Van Den Berg, "Generative AI and Educators: Partnering in using open digital content for transforming education," *Open Praxis*, vol. 16, no. 2, pp. 130–141, Jan. 2024, Available: https://doi.org/10.55982/openpraxis.16.2.640
[24] Z. D. Li and T. V. Harper, "Augmented Reality Applications in K-12 Education," *Educational Technology Studies*, vol. 8, no. 5, pp. 233–240, 2022.
[25] White House Office of Science and Technology Policy, "Blueprint for an AI Bill of Rights: Making Automated Systems Work for the American People," *White House OSTP*, Oct. 2022. [Online]. Available: www.whitehouse.gov/ostp/ai-bill-of-rights
[26] EDUCAUSE Review, "10 Ways Artificial Intelligence Is Transforming Instructional Design," *EDUCAUSE*, 2024. [Online]. Available: https://er.educause.edu/articles/2024/10/10-ways-artificial-intelligence-is-transforming-instructional-design
[27] HowWithAI, "AI In Teaching: Transforming Traditional Teaching Methods With Artificial Intelligence," 2024. [Online]. Available: www.howwithai.com/articles/ai-teaching-transforming-traditional-methods
[28] STEM Education Guide, "How to Use AI for Educators: Enhancing Classroom Learning," *STEM Education Guide*, 2024. [Online]. Available: www.stemeducationguide.com/how-to-use-ai-for-educators

5

Artificial Intelligence: A Digital Architect for Curriculum Development

Shreya Tyagi, Taranjeet Duggal and Srishti Singh Kaira

5.1 Introduction

Conversations about AI take us back to the year 1956 – to the Dartmouth Conference in which the term "artificial intelligence" was coined by John McCarthy (2U WordPress, 2023). Although the AI journey in India was started in 1960 by Professor H.N. Mahabala, a concrete foundation of AI in research and education was not built till the 21st century when institutes like IIT Kanpur and Indian Statistical Institute started making a significant contribution to the field (Kitcbe-Blog, 2023; *The Rise and Roar of AI in India: A Transformative Journey*, n.d.).

In a growing country like India, we can see rapid advancements and acceptance toward AI, especially in the fields of education and research. Here, the ideal student–teacher ratio should be 30:1 in the primary level and 35:1 in the upper primary levels (Student–Teacher Ratio, n.d.). This means one teacher for 30 students; the numbers here are a forthright example of increasing pressure on educators while maintaining efficiency in their work. To alleviate the pressure and maintain a healthy work–life balance, AI has been integrated into the area of curriculum design. It not only reduces stress for educators but also promotes personalized learning styles and a fair grading system for students (Evanick, 2024).

Every student is different, every mind is different; hence, it is very difficult for one teacher to adjust to the requirements of every student. Therefore, AI introduces personalized learning pathways that can be created to cater for the needs of every child, not just those who are neurotypical but also those who are neuroatypical. These pathways analyse assessments, feedback, and student interaction (Rekha et al., 2024), thereby improving student engagement and achieving academic goals through balancing the interest areas (Onesi-Ozigagun et al., 2024).

DOI: 10.1201/9781003560814-5

Looking at the gaps and the need of the hour, this chapter will be a walking guide for educators on how to manoeuvre AI in curriculum design while keeping ethical considerations in mind. The chapter will also focus on the use of generative AI in education, with a section focusing on children with special needs. The ending of the chapter would be incomplete without explaining the role of educators in an AI-driven world. The section focuses on the need for training in the field and using AI as a supplement and not a replacement. The chapter not only addresses the use of AI but also discusses when not to use AI and how much AI should be used.

5.2 Generating Efficiency Using AI

AI is efficient in executing tasks. One of the major features of AI is producing efficient work. This efficiency can be achieved using technologies like enhancing creativity, streamlined grading systems, and individualized learning paths (Hashem & Hakeem, 2024). Apart from this, AI also helps educators update courses according to recent research, making the environment more dynamic for learning (Rufrano & Yeung, 2023). Out of the many key features of AI, this section aims to highlight those specifically important for curriculum development:

i) *Automatic content generation:* This refers to the yielding of content such as quizzes, assignments, or homework based on individual learning styles. This technology can not only create new content but also automatically yield data for routine tasks. For example, a teacher can use this to create an assignment every week without worrying about data gathering or finding information. It helps in the reduction of the administrative work of the educator and helps them to focus on the understanding of each student (Arellano, 2024). One such tool is IBM Watson, which is significantly working on managing repetitive and administrative tasks with the help of IBM Watson Orchestrate and Robotic Process Automation (RPA) (*Platform – IBM Watson Orchestrate | IBM*, n.d.). Educators can create skills and assign them to Watson, which can then use them to manage schedules or analyse the trend in students' performance and help educators in creating a working plan to address those gaps (Chirag, 2024)

ii) *Optimization of class schedules:* AI's ability to allocate resources efficiently by tweaking class timetables and making chatbots accessible is evolving the education industry by initiating individualized learning maps. By accessing information like class availability, teacher timetables, and student preferences, AI can manage conflicts and promote a solution-focused approach. OptaPlanner is an AI-based tool that critically analyses complex scheduling problems while keeping in mind physical constraints like classroom capacity or teacher availability (*School Timetabling*, n.d.). On the other hand, StatBot is one of the chatbots utilized in the academic sector, acting as an immediate support to educators and students that monitors administrative and conceptual tasks and provides personalized assistance to students (Karunarathne et al., 2024). These chatbots are trained to be inclusive by natural language processing (NLP) and machine learning algorithms that help AI understand the specific queries of every student (Chakraborty et al., 2023). The pedagogical principles of theories like Cognitive Load Theory and Adaptive Learning Theory also align with the implementation of chatbot Learning Management Systems (LMS) to promote comprehension and engagement (Mungai et al., 2024).

iii) *Data-driven curriculum adjustments:* AI is trained to continuously monitor data that it gathers from the assessments, attendance, and performance of the student. These data are recorded in real time. This mechanism is the continuous data analysis that helps AI prompt educators when there is a need for adjustment in the curriculum (Stefanic, 2024). With refined technologies such as natural language processing and machine learning, analysing large data sets is seamlessly managed by AI (Pusporini & Nurdiyanto, 2024). Similarly, another way through which AI adjusts curricula is by using predictive analytics. Here, AI employs machine learning to predict when a student may fall behind based on past trends. It signals the educator to give special attention to that student. This not only helps in assessing the weaknesses of the child but also intervenes before it is too late and makes necessary changes in the curriculum (Theodotou, 2024).

We all know that creating personalized learning pathways is what AI does the best (*8 Transformative Use Cases of AI for Curriculum Design*, 2024), but apart from this it also provides instantaneous feedback to students, enabling educators to monitor their classroom's progress and endorsing students to improve and understand their weaknesses (Hamdi, 2024; Stefanic, 2024). Generative AI is revolutionizing the field of education by automatically generating creative content such as quizzes, lessons, and even entire modules. One such tool that can not only help educators in deflating the pressure but also provide engaging content for students is Smart Sparrow. In the wide array of AI tools, Smart Sparrow is one such tool that can incorporate various multimedia elements like simulations or video or audio content to make a whole interactive module for students (*Smart Sparrow Adaptive Tutorial | Teaching Technology ToolKit*, n.d.).

We know from the above discussion that as important as AI is for education, we cannot neglect the role of human teachers when it comes to decision-making in critical situations. AI might produce data-driven inferences that can elevate curriculum development and teaching methods, but it is pertinent to ensure that educators perpetuate their role as leaders and authority figures in the classroom. AI is not a substitute for a mentor's judgement but a reassuring tool that can be used for educational motives. To make classrooms more dynamic, ensuring that every need is catered to, educators should be able to understand AI and implement these insights in day-to-day teaching. One of the very prevalent risks of using AI can be over-reliance of educators, which could cloud their professional judgement. It is crucial to foster a culture where teachers can critically assess AI results rather than unquestioningly accept them (Ghanrawi et al., 2023).

5.3 Individualized Learning Paths

AI-enabled learning changes the dynamics of a classroom from a traditional homogenized setup to a more personalized learning environment. Below are a few advantages of using AI-based learning in the classroom:

- *Dynamic Adjustment* – Dynamic adjustment as per the skill levels and learning goals of the students (C et al., 2024). This personalization is achieved through the system's ability to analyse patterns in student learning preferences and outcomes and predict and prescribe tailored learning materials while also reducing the human effort in designing effective learning paths for varied learners (Amal et al., 2024; Widono et al., 2024). It provides the space to adjust the pace of learning and difficulty of the content, as per the needs, interests, abilities, and existing progress of the learners.
- *User Engagement* – The aspect of user engagement is quite high in an AI-based classroom. It involves interactive tests, performance feedback, and specific video recommendations (C et al., 2024). The provision of real-time feedback allows learners to efficiently adapt their strategies accordingly, boosts confidence, and keeps them engaged. One such AI-driven platform is *Modulo*, which integrates smartly chosen YouTube videos, tutorials, and other supplementary materials, allowing students to choose from a range of their subjects' content (Amal et al., 2024). It also makes use of an iterative waterfall approach, allowing continuous feedback and adaptation (Amal et al., 2024).

- *Accessibility and Inclusivity in Education* – A major advantage of AI in academics is that it improves accessibility for all kinds of learners, forming a more inclusive environment. For instance, AI-based translations and captioning have made it easier for learners with hearing impairments or those with multilingual backgrounds to access learning materials as per their needs (Adeleye et al., 2024). For learners with neuromuscular conditions, an eye-tracking AI-based system is of utmost importance. This innovative technology helps in learning by tracking the direction of the person's vision, how long the gaze lasts and whether they focus on a specific object or stimuli (Valencia-Londoño et al., 2024). This has been tested to be vital in reducing communication and physical barriers in a classroom setting. Similarly, AI-based simulation tools can help learners on the autism spectrum or those with ADHD by engaging them in practice-based assignments and hands-on training and providing the opportunity to practice complex analytical skills (Ramasamy & Lee, 2024).
- *Collaborative Learning* – AI-based e-collaborative tools open up the possibilities of online group discussion platforms, simultaneously working on projects irrespective of where the learners are in terms of their locations (Kim et al., 2022; Adelaide et al., 2024). Due to the availability of multidisciplinary sections in an artificially intelligent set-up, it becomes easier for learners to gain knowledge, collaborate with people from different subject areas, and even work on socially relevant problems (Kim et al., 2022). Intelligent Tutoring Supported Collaborative Learning (ITSCL), MOOCs, and Computer Supported Collaborative Learning (CSCL) help in connecting learners from various backgrounds, aiding in building peer learning circles and also improving tutor–learner interactions, leading to a comprehensive learning environment (Haq et al., 2020; Kim et al., 2022).
- *Performance Tracking and Assessments* – AI, not just for students but for educators too, has proven to be a time-efficient and supportive tool. It helps in the timely understanding and prediction of learning difficulties and planning of necessary interventions, highlighting areas for improvement, and promotes quick assessment and effective decision-making (Halagatti et al., 2023; Bulut et al., 2024). Automated grading, personalized feedback, enhanced accuracy and fairness, reduced teacher load, and timely feedback are some of the advantages of using AI-based assessment methods (Bulut et al., 2024; Lyanda et al., 2024; Kayalı & Balat, 2024). Over the last few years, Natural Language Processing (NLP)-based technologies have been used to analyse the contents of discourses that aid in assessing the learning outcomes of diverse learners (Tan et al., 2022).

5.4 AI in Real Life

The traditional education system is experiencing a pivot in its structure, where conventional teaching styles are being replaced by integrating AI into learning pathways and curriculum design. However, attempts to redesign the curriculum are not new; although AI has been misinterpreted, people have been trying to integrate it for a long time. For example, in the US, Common Core States Standards aimed to develop skill consistency. But it turned out to be a total failure because they assumed unified progression for everyone (Sharma, 2024). Learning from these mistakes, some schools and universities have worked on improving the robustness of AI and machine learning capabilities and applied them to solve first-hand problems that they were experiencing. Below are some case studies to delve into real-life applications of AI across the world:

CASE STUDY 1

Problem: The University of Sydney was facing the challenge of providing personalized learning to diverse students across a very large campus. They aimed to deliver individualized content as a better approach towards education.

Solution: They used software called Smart Sparrow that helped teachers to design individualized learning pathways and, accordingly, shape coursework and adapt immediately according to the calibre of the students.

Key Impact: Adopting Smart Sparrow enhanced performance and made students more interactive by interacting with each other more and participating in classroom activities/discussions more. The teachers also discovered a new understanding of the learning patterns of students, allowing them to adopt new strategies.

CASE STUDY 2

Problem: Harris Federation, a school in the United Kingdom, was facing challenges in managing the workload of teachers while also maintaining the accessibility of the curriculum to students of diverse cultures and backgrounds.

Solution: To overcome these challenges, Harris Federation used ChatGPT and Microsoft Live. ChatGPT changes the text according to the different age groups of students, making it more accessible, and Microsoft Live translated content in real time, providing live subtitles from different language backgrounds.

Key Impact: The time spent on completing administrative work of designing curriculums was significantly reduced. This, allowed teachers to use that time in building student–teacher interactions, thereby improving the learning experience.

CASE STUDY 3

Problem: Mentors at Modern School, India, were burdened with the prolonged task of assessing long essays, which restricted them from engaging with students in interactive and creative teaching methods.

Solution: Modern School implemented an AI-enabled automated essay-scoring system. The system provides scores by checking the grammar, quality of content, argumentation, and coherence. These scores are accompanied with detailed feedback on how to improve areas of concern for each student.

Key Impact: Teachers were able to utilize their time in using other creative methods of teaching, making education a fun learning experience. Apart from this, students also benefitted from quick feedback, enabling timely revisions and learning (DigitalDefynd, 2024).

India, though late, has entered the world of AI. It is ready to address barriers, taboos, and challenges around incorporating AI in education. With balanced AI–teacher interaction, schools can magnify the learning experience so that no student feels left out.

5.5 Ethical Considerations and Challenges

With the increase in illustrious use of AI in various fields, whether science, social science, or education, it is a major responsibility to train educators and apprentices in the correct use of AI. AI works on algorithms developed by humans; therefore, it is of utmost importance to check for the security and privacy of students. To ensure these educational institutions implement robust parameters so that AI can only be used for educational benefits rather than harming students. The following is a list of measures that can be taken by schools/colleges to ensure safety.

5.5.1 Data Privacy and Security

- *Data Specification:* Educational institutions should only focus on collecting specific and relevant details necessary to achieve the personalized goals of students. Through this, educators can minimize the risk of storing large amounts of confidential data (*Artificial Intelligence in Schools: Privacy and Security Consideration*, n.d.).
- *Transparency and Consent:* To build trust and resolve doubts around AI, educators need to be honest about the information they give to parents and guardians. One should obtain informed consent from parents and students before collecting or using their data (Merod, 2024).
- *Adhering to Privacy Laws:* Educational institutes must adhere to the legal framework governing the concerns around the use of AI for students. The Digital Personal Data Protection Act (DPDT) 2023 talks about the requirement of consent, data minimization, and protection against misuse of data (Lohchab, 2024). Apart from this, one of the existing laws in India is the Information Technology Act, 2000. This discusses the regulations and penalties related to data breaching offences and cybersecurity (Gupta, 2024).
- *Vendor Vetting and Active Steps towards Security:* Before hiring a third-party vendor, schools/colleges must have a strict protocol to vet the vendor by reviewing their privacy policies and whether they comply with the laws of data privacy. In addition, schools must check the previous records of the vendor before signing in (Merod, 2024). On the part of the school, they should have robust security, ensuring data protection, access to controls, and constant security audits to protect students from unauthorized data breaches (*Artificial Intelligence in Schools: Privacy and Security Considerations*, n.d.).

5.6 Biases in AI

AI systems may be impeccable when providing education, but they can unwittingly perpetuate the biases stored in their training data. Surfacing biases can impact the learning of the student, especially in culturally driven areas. It has been observed that biases have an egregious impact on students when it comes to being culturally sensitive. If the AI platform being used is made according to Western culture, then students from other cultures might feel excluded or even offended by certain remarks about their culture. This is not just about the feelings of the student but also the learning outcomes; such students won't be able to learn about diversity, and it may diminish their motivation to study (Francisco, 2023; Karpouzis, 2024). Biases in AI can promote skewed stereotypes about various cultures and values. This can be seen as reinforcing ideas about a particular gender or ethnicity in a negative light, which can, in turn, develop into biased opinions of students. Therefore, with limited exposure to diverse data and increased emphasis on particular cultures, the self-esteem of the students can be affected negatively (*Impact of AI on Content Diversity and Cultural Sensitivity*, 2024; Liu, 2024). A study published in the academic *Journal of Digital Medicine* revealed that Large Language Models (LLM), which are a pertinent part of generative AI when implemented in healthcare systems, have the potential to perpetuate race-based medicines (*Five Strategies to Mitigate Bias When Implementing Generative AI*, 2024). Race-based medicine is a flawed medical practice that treats individuals based on their race, which can be biased and discriminatory (Dutchen, 2024).

5.7 Strategies to Mitigate Biases

GenAI poses some great risks of social biases, which should be taken care of by developers and those who deploy them. In a survey by Telus, 32% of people said that they had missed opportunities with jobs and experienced financial loss due to bias within AI algorithms, and a further 40% believed that companies that use AI are not doing enough to protect their users from biased AI algorithms (*Five Strategies to Mitigate Bias When Implementing Generative AI*, 2024). Since AI is not just a mere tool to enhance efficiency but also a platform that affects its user's life, it is of absolute importance to regulate the use of AI and strategize accordingly.

- Using a data-centric approach is one of the ways forward. Ensuring that AI produces diverse and culturally rich information along with the fair generation of data allows students to gain access to unbiased information (González-Sendino et al., 2024; Mishra et al., 2024).
- Using fair awareness in-process techniques directly during the model training, effectively reduces disparate impact and statistical parity differences (Soni, 2024).
- After implementing AI systems, institutions must ensure robust evaluation parameters (Mishra et al., 2024).

- Techniques like reweighing and adversarial debiasing should enhance transparency, especially in chatbots (Gupta et al., 2024).

Apart from these four points that highlight some crucial ways to mitigate biases, there is a five-step process that gives it a structure and systematic approach rather than being scattered.

i) *Prioritizing the Need to Mitigate Biases:* Form a strong leadership team that focuses on reducing biases in AI. An organization should support and fund initiatives that are working towards the same goal. People should invest in ethically built AI systems such as "Constitutional AI" installed within Claude, the largest LLM by Anthropic, which stops Claude from doing anything unethical.

ii) *Mandate Bias-Mitigation Policy:* Create clear written policies that are looked over by a governance team to balance the use of AI.

iii) *Use Various Training Data:* While training an AI model, one must use diverse and reliable sources to train AI in covering a large number of cultural biases. Using visualizations helps detect and correct these biases while anonymising sensitive information.

iv) *Fine-Tune with Diversity:* To make AI more cohesive with cultural nuances, continuous human feedback, also known as reinforcement learning from human feedback (RLHF), and red teaming the LLM, which is a method that consciously attacks the AI system through prompts that uncover its flaws and vulnerabilities, are used.

v) *Keep a Check on the AI Model:* The way AI systems react during training can be different from when exposed to the real world. This is because of a phenomenon known as data shift, which occurs due to different prompts being used in the real world as compared to what it was trained in. To prevent this, continuously monitor the AI model and keep fine-tuning as it becomes more exposed to the real world (*Five Strategies to Mitigate Bias When Implementing Generative AI*, 2024).

5.8 Role of Teachers in an AI-Dominant World

As AI becomes more prevalent in education, concerns often arise about whether it could replace teachers (Wang et al., 2024). However, AI is a powerful supplement and not a replacement (Yan, 2023). Educators are not only responsible for delivering content but also for providing emotional, social, and ethical support – roles that AI cannot fulfil. In an AI-driven world, teachers will continue to play a noteworthy role in shaping students' cognitive, emotional, and social development. According to Seo et al. (2021), the collaboration between AI and educators holds the potential to craft a more goal-driven and supportive learning environment. Though optimizing tasks such as grading, data analysis, and content delivery is its forte, it lacks the human touch required to connect with students on a deeper level (Deloitte, 2021). Teachers remain irreplaceable when it comes to analysing data, individualizing learning, and addressing the needs of students. Beyond academics, teachers act as mentors, helping students develop critical thinking, problem-solving abilities, resilience, and emotional intelligence. By now, from the above discussion, we can see that the judicial use of AI can help teachers, but it cannot replicate the interpersonal relationships that foster a safe, inclusive, and motivating learning environment between the educator and the student. For AI to be an effective educational tool, teachers must be trained to use it skilfully. Developing specialized skills should not be the only agenda of employee development programs, it should also focus on how to integrate AI into teaching strategies that enhance student outcomes. Making teachers aware of AI can ensure their critical role in student development. As AI takes over daily tasks, we will be able to see a shift in the role of teachers from being knowledge transmitters to facilitators, guides, and mentors (Ghamrawi et al., 2023). Through this, teachers can focus on developing soft skills, such as creativity, emotional intelligence, collaboration, and ethical reasoning, which are increasingly important in a world where automation is prevalent. AI systems may handle content delivery and student performance tracking, but teachers will be responsible for guiding students through deeper learning experiences, encouraging inquiry-based learning, and supporting their emotional and social growth (Celik et al., 2022). Moreover, the ability to build strong student–teacher relationships through empathy and understanding will become even more crucial in this new era. According to the World Economic Forum, as AI continues to advance, the human facets of teaching – like compassion, patience, and the ability to inspire students – will grow in importance (*How AI Can Accelerate Students' Holistic Development and Make Teaching More Fulfilling*, 2024). Teachers who master these skills will be those most successful in equipping students for the intricate challenges of the future. In addition, teachers will play a vital role in educating students on the ethical use of AI, fostering discussions on its societal impact, and ensuring that students become responsible digital citizens (Celik et al., 2022).

With numerous benefits in place comes some negative impact as well:

- From the previous sections, we know that biases in AI are a major concern; therefore, if there is no governing team, one won't be able to combat the biases in the education system as well (Akgun & Greenhow, 2021).

An example to explain this could be an Indian boy being considered weak in studies as compared to a white boy.

- Teachers or even students can become over-dependent on technology to deliver knowledge. This may hamper the learning process in the case of technical failures.
- With a growing individualistic culture, the only bonds some people have are through their parents, teachers, and limited numbers of friends. With excessive use of AI, human interaction will become even more limited, making it difficult for children to be more emotionally resilient (Vernersson, 2025).
- The last and most discussed disadvantage of AI in academics is cheating, as students can use AI to generate work that looks original (Vernersson, 2025).

All in all, AI is not a tool for replacement but a way to make a synchronization between technology, teachers, and students, thereby improving their relationship and education quality.

5.9 Conclusion

To conclude, the amalgamation of AI into curriculum development is changing the education landscape by offering innovative answers to questions about the future of the education industry. AI greatly increases efficiency by automating day-to-day tasks, such as quiz and assessment creation, allowing educators to devote more time to teaching and interacting with students. AI-powered tools help provide material that is especially curated to take care of the needs of every student, assisting in bridging learning gaps and supporting a wide variety of learning styles.

However, as AI transforms the educational landscape, important ethical questions also come into the picture. Issues such as data privacy, bias in algorithms, and a change in the role of teachers must be addressed to ensure that AI is used responsibly. The amalgamation of educators and policymakers is needed to create a balanced system between technology and human aspects of teaching to make sure that AI tools support the educator's role rather than replace them.

AI is not only a way to create a more dynamic and inclusive environment but, if implemented judicially, can prepare students to meet the increasing demands of technological advancement with support towards curriculum development. Balancing these changes with ethical considerations will be a door to maximize the benefits of AI in education.

References

2U Wordpress. (2023, October 27). *History of AI: Timeline and the Future | Maryville Online*. Maryville University Online. Retrieved February 27, 2025, from https://online.maryville.edu/blog/history-of-ai/

Adeleye, N. O. O., Eden, N. C. A., & Adeniyi, N. I. S. (2024). Innovative teaching methodologies in the era of artificial intelligence: A review of inclusive educational practices. *World Journal of Advanced Engineering Technology and Sciences*, 11(2), 069–079. https://doi.org/10.30574/wjaets.2024.11.2.0091

Akgun, S., & Greenhow, C. (2021). Artificial intelligence in education: Addressing ethical challenges in K-12 settings. *AI And Ethics*, *2*(3), 431–440. https://doi.org/10.1007/s43681-021-00096-7

Amal, S. F., Saiid, I. A., & Mansor, H. (2024). An empirical study for the dynamic and personalised learning experience of the AI course Generator. *International Journal on Perceptive and Cognitive Computing*, *10*(2), 23–30. https://doi.org/10.31436/ijpcc.v10i2.483

Arellano, B. (2024, July 9). *Using AI for Curriculum Design + 21 AI Prompt Suggestions – Fly Plugins*. Fly Plugins. https://flyplugins.com/ai-for-curriculum-design/

Artificial Intelligence in Schools: Privacy and Security Considerations. (n.d.). New America. www.newamerica.org/oti/blog/artificial-intelligence-in-schools-privacy-and-security-considerations/

Bist, A. (2024, October 18). 8 Transformative Use Cases of AI for Curriculum Design. (2024, October 18). www.webuters.com/use-cases-of-ai-for-curriculum-design

Bulut, O., Beiting-Parrish, M., Casabianca, J. M., Slater, S. C., Jiao, H., Song, D., Ormerod, C. M., Fabiyi, D. G., Ivan, R., Walsh, C., Rios, O., Wilson, J., Yildirim-Erbasli, S. N., Wongvorachan, T., Liu, J. X., Tan, B., & Morilova, P. (2024). The rise of artificial intelligence in educational measurement: opportunities and ethical challenges. *arXiv (Cornell University)*. https://doi.org/10.48550/arxiv.2406.18900

C, N. M., S, K., R, H. M., D, K. M., & M, K. (2024). AI enabled Personalized Learning platform. *International Journal for Multidisciplinary Research*, *6*(3). https://doi.org/10.36948/ijfmr.2024.v06i03.23253

Celik, I., Dindar, M., Muukkonen, H., & Järvelä, S. (2022). The promises and challenges of artificial intelligence for teachers: A systematic review of research. *TechTrends*, *66*(4), 616–630. https://doi.org/10.1007/s11528-022-00715-y

Chakraborty, C., Pal, S., Bhattacharya, M., Dash, S., & Lee, S. (2023). Overview of Chatbots with special emphasis on artificial intelligence-enabled ChatGPT in medical science. *Frontiers in Artificial Intelligence*, *6*. https://doi.org/10.3389/frai.2023.1237704

Chirag. (2024, October 15). 12 ways AI in education is transforming the industry. *Appinventiv*. https://appinventiv.com/blog/artificial-intelligence-in-education/

Chiu, T. K. F. (2021). A holistic approach to the design of Artificial Intelligence (AI) education for K-12 schools. *TechTrends*, *65*(5), 796–807. https://doi.org/10.1007/s11528-021-00637-1

Deloitte. (2021). Generative AI and higher education. In *Higher Education* [Report]. www2.deloitte.com/content/dam/Deloitte/ca/Documents/deloitte-analytics/ca-generative-ai-and-higher-education-en.pdf

DigitalDefynd, T. (2024, September 6). Use of AI in Schools [25 case studies] [2025] – DigitalDefynd. DigitalDefynd. Retrieved February 28, 2025, from https://digitaldefynd.com/IQ/ai-in-schools-case-studies/

Dutchen, B. S. (2024, December 12). Field correction. *Harvard Medicine Magazine*. https://magazine.hms.harvard.edu/articles/field-correction#:~:text=For%20centuries%2C%20race%2Dbased%20medicine,ranges%20far%20beyond%20pain%20management.

Elov, B., Samatov, M., Gayibova, N., Samadova, N., Qodirova, M., & Amirovich, S. (2024). A mantle of Chatbots in the place of pedagogy and learning to generate tools to provide assistance using A-I. In Sanjoy Das, Marcin Paprzycki, Ankush Ghosh, Monica Bianchini (eds.). *2024 4th International Conference on Advance Computing and Innovative Technologies in Engineering (ICACITE)*. Springer Singapore, (Vol. 154, pp. 1075–1079). https://doi.org/10.1109/icacite60783.2024.10617382

Evanick, J., EdD. (2024, June 28). *Integrating AI In Curriculum Design: A Comprehensive Guide For Educators*. eLearning Industry. https://elearningindustry.com/integrating-ai-in-curriculum-design-a-comprehensive-guide-for-educators

Five strategies to Mitigate bias when Implementing Generative AI. (2024, March 14). Telus Digital. Retrieved February 27, 2025, from www.telusdigital.com/insights/ai-data/article/mitigating-genai-bias

Francisco. (2023, September 18). Navigating Cultural Sensitivities with AI-Assisted Teaching – Teachflow.AI. *Teachflow.AI*. https://teachflow.ai/navigating-cultural-sensitivities-with-ai-assisted-teaching/

Ghamrawi, N., Shal, T., & Ghamrawi, N. A. (2023). Exploring the impact of AI on teacher leadership: Regressing or expanding? *Education and Information Technologies*, *29*(7), 8415–8433. https://doi.org/10.1007/s10639-023-12174-w

González-Sendino, R., Serrano, E., & Bajo, J. (2024). Mitigating bias in artificial intelligence: Fair data generation via causal models for transparent and explainable decision-making. *Future Generation Computer Systems*, *155*, 384–401. https://doi.org/10.1016/j.future.2024.02.023

Gupta, R. P. (2024, July 15). *How to Harness AI in Schools: Opportunities and Challenges ahead*. IndiaAI. Retrieved June 13, 2025, from https://indiaai.gov.in/article/how-to-harness-ai-in-schools-opportunities-and-challenges-ahead

Gupta, S., Kundu, R., Deo, A. K. A., Patnaik, M., Kundu, T., & Dehury, M. K. (2024). Enhancing transparency and mitigating bias in large language models' responses with sophistication. In Sanjoy Das, Marcin Paprzycki, Ankush Ghosh, Monica Bianchini (eds.). *2024 IEEE International Conference on Information Technology, Electronics and Intelligent Communication Systems (ICITEICS)*. Springer Singapore, (Vol. 30, pp. 1–6). https://doi.org/10.1109/iciteics61368.2024.10625096

Halagatti, M., Gadag, S., Mahantshetti, S., Hiremath, C. V., Tharkude, D., & Banakar, V. (2023). Artificial intelligence: the new tool of disruption in educational performance assessment. In *Contemporary studies in economic and financial analysis* (pp. 261–287). https://doi.org/10.1108/s1569-37592023000110a014

Hamdi, M. (2024). How AI is transforming and shaping the future of education. In *2024 IEEE 28th International Conference on Intelligent Engineering Systems (INES)*. IEEE, (pp. 000115–000116). https://doi.org/10.1109/ines63318.2024.10629089

Hashem, O. A., & Hakeem, M. B. (2024). Design education methodology using AI. *Deleted Journal*, *3*(1), 121–135. https://doi.org/10.55554/2785-9649.1030

Haq, I. U., Anwar, A., Basharat, I., & Sultan, K. (2020). Intelligent Tutoring Supported Collaborative Learning (ITSCL): A Hybrid Framework. *International Journal of Advanced Computer Science and Applications*, 11(8). https://doi.org/10.14569/ijacsa.2020.0110866

How AI can accelerate students' holistic development and make teaching more fulfilling. (2024, September 10). World Economic Forum. www.weforum.org/agenda/2023/05/ai-accelerate-students-holistic-development-teaching-fulfilling/

Impact of AI on content diversity and cultural sensitivity. (2024, June 20). FasterCapital. https://fastercapital.com/content/Impact-of-ai-on-content-diversity-and-cultural-sensitivity.html

Karpouzis, K. (2024). Plato's shadows in the digital cave: Controlling cultural bias in generative AI. *Electronics, 13*(8), 1457. https://doi.org/10.3390/electronics13081457

Karunarathne, W., Paladino, A., Selman, C., Nagy, K., Sajitos, L., & Kishore, S. (2024). The StatBot. *Pacific Journal of Technology Enhanced Learning, 6*(1), 20–21. https://doi.org/10.24135/pjtel.v6i1.193

Kitcbe-Blog. (2023, December 6). *India's First Experience With Ai – Karpagam Institute*. Karpagam Institute of Technology. https://karpagamtech.ac.in/indias-ai-journey/

Liu, Y. (2024). Unveiling bias in artificial intelligence: Exploring causes and strategies for mitigation. *Applied and Computational Engineering, 76*(1), 124–133. https://doi.org/10.54254/2755-2721/76/20240576

Lohchab, H. (2024, September 20). India's AI strides run into privacy law headwinds. *The Economic Times*. https://economictimes.indiatimes.com/tech/artificial-intelligence/indias-ai-strides-run-into-privacy-law-headwinds/articleshow/113500325.cms

Lyanda, J. N., Owidi, S. O., & Simiyu, A. M. (2024). Rethinking Higher Education Teaching and Assessment In-Line with AI Innovations: A Systematic Review and Meta-Analysis. *African Journal of Empirical Research, 5*(3), 325–335. https://doi.org/10.51867/ajernet.5.3.30

Merod, A. (2024, May 9). How schools can vet AI tools to protect student data. *K-12 Dive*. www.k12dive.com/news/AI-vetting-ed-tech-schools-checklist/715623/

Mishra, N. I., Kashyap, N. V., Yadav, N. N., & Pahwa, N. D. R. (2024). Harmonizing intelligence: A holistic approach to bias mitigation in Artificial Intelligence (AI). *Deleted Journal, 2*(7), 1978–1985. https://doi.org/10.47392/irjaeh.2024.0270

Mungai, B. K., Omieno, P. K. K., Egessa, M., & Manyara, P. N. (2024). AI Chatbots in LMS: A pedagogical review of cognitive, constructivist, and adaptive principles. *Engineering and Technology Journal, 9*(8), 4709-4715. https://doi.org/10.47191/etj/v9i08.15

Onesi-Ozigagun, N. O., Ololade, N. Y. J., Eyo-Udo, N. N. L., & Ogundipe, N. D. O. (2024). Revolutionizing education through AI: A comprehensive review of enhancing learning experiences. *International Journal of Applied Research in Social Sciences, 6*(4), 589–607. https://doi.org/10.51594/ijarss.v6i4.1011

Petersen, C. (2024). Visible learning: The sequel: A synthesis of over 2,100 meta-analyses relating to achievement Visible learning: The sequel: A synthesis of over 2,100 meta-analyses relating to achievement, by John Hattie, London, Routledge, 2023, 510 pp., £18.39(Paperback), ISBN: 9781032462035. *Educational Review*, 77(1), 1–2. https://doi.org/10.1080/00131911.2024.2302742

Platform – IBM Watson Orchestrate | IBM. (n.d.). Watsonx Orchestrate: AI agents for business, www.ibm.com/products/watsonx-orchestrate/platform

Pusporini, W., & Nurdiyanto, H. (2024). Utilization of artificial intelligence in outcome-based curriculum evaluation and development. *Journal of Research in Social Science and Humanities*, *4*(2), 132–134. https://doi.org/10.47679/jrssh.v4i1.109

Rekha, K., Gopal, K., Satheeskumar, D., Anand, U. A., Doss, D. S. S., & Elayaperumal, S. (2024). Ai-powered personalized learning system design: Student engagement and performance tracking system. In Sanjoy Das, Marcin Paprzycki, Ankush Ghosh, Monica Bianchini (eds.). *2024 4th International Conference on Advance Computing and Innovative Technologies in Engineering (ICACITE)*. Springer Singapore, (Vol. 28, pp. 1125–1130). https://doi.org/10.1109/icacite60783.2024.10617155

R.K, R. (2023, October 8). *The Rise and Roar of AI in India: A Transformative Journey*. (n.d.). INDIAai. https://indiaai.gov.in/article/the-rise-and-roar-of-ai-in-india-a-transformative-journey

Rufrano, M. A., & Yeung, J. (2023). Digital pedagogy for the present: An artificial intelligence methodology for curriculum development. In Josep Domenech, David Menéndez Álvarez-Hevia, Alicia Martínez-Varea, Rosa M. Llácer-Iglesias & Domenico Brunetto (eds.). *9th International Conference on Higher Education Advances (HEAd'23)*. Universitat Politècnica de València, (pp. 757–764). https://doi.org/10.4995/head23.2023.16175

School timetabling. (n.d.). OptaPlanner. www.optaplanner.org/learn/useCases/schoolTimetabling.html

Seo, K., Tang, J., Roll, I., Fels, S., & Yoon, D. (2021). The impact of artificial intelligence on learner–instructor interaction in online learning. *International Journal of Educational Technology in Higher Education*, *18*(1). https://doi.org/10.1186/s41239-021-00292-9

Sharma, K. (2024, February 9). AI as a guide for developing better curriculum and learning paths. *Times of India Blog*. https://timesofindia.indiatimes.com/blogs/ai-musings/ai-as-a-guide-for-developing-better-curriculum-and-learning-paths/

Smart Sparrow Adaptive Tutorial | Teaching Technology Toolkit. (n.d.). UNSW Staff Teaching Gateway. www.teaching.unsw.edu.au/digital-assessment-toolkit/exemplar-1

Soni, V. (2024). Bias detection and mitigation in AI-driven target marketing: Exploring fairness in automated consumer profiling. *International Journal of Innovative Science and Research Technology (IJISRT)*, 2574–2584. https://doi.org/10.38124/ijisrt/ijisrt24may2203

Stefanic, D. (2024, July 30). *Revolutionize Feedback: AI-Powered Real-Time Learning Insights*. Hyperspacemv – the Metaverse for Business Platform. https://hyperspace.mv/revolutionize-feedback-ai-powered-real-time-learning-insights/

Student–Teacher Ratio. (n.d.). Student-Teacher Ratio. https://pib.gov.in/newsite/PrintRelease.aspx?relid=158326

Theodotou, M. (2024, January 12). *2024 Trends: AI-Powered Learning Analytics*. eLearning Industry. https://elearningindustry.com/2024-trends-ai-powered-learning-analytics

undefined. (n.d.). INDIAai. https://indiaai.gov.in/article/how-to-harness-ai-in-schools-opportunities-and-challenges-ahead

Valencia-Londoño, P. A., Cardona-Rodas, H., & Jiménez-Builes, J. A. (2024). An Educational Inclusion Model for Adults with Diverse Neuromuscular Conditions through the use of an Artificial Intelligence Algorithm. *International Journal of Emerging Technologies in Learning (iJET)*, *19*(05), 35–56. https://doi.org/10.3991/ijet.v19i05.46619

Vernersson, N. (2025, January 13). 10 advantages & disadvantages of artificial intelligence in education. *Digiexam*. Retrieved February 27, 2025, from www.digiexam.com/blog/advantages-and-disadvantages-of-artificial-intelligence-in-education

Wang, S., Wang, F., Zhu, Z., Wang, J., Tran, T., & Du, Z. (2024). Artificial intelligence in education: A systematic literature review. *Expert Systems With Applications*, *252*, 124167. https://doi.org/10.1016/j.eswa.2024.124167

Widono, S., Mulyadi, N., Saddhono, K., Nurhasanah, F., Nugraheni, A. S. C., & Legowo, B. (2024). *A Strategic design of personalized based learning system for improving the experience of outcome based education* (Vol. 15, pp. 1149–1154). 4th International Conference on Advance Computing and Innovative Technologies in Engineering (ICACITE). https://doi.org/10.1109/icacite60783.2024.10616811

Yan, D. (2023). Visible learning: The sequel: A synthesis of over 2,100 meta-analyses relating to achievement Visible learning: The sequel: A synthesis of over 2,100 meta-analyses relating to achievement, by John Hattie, Abingdon, Routledge, 2023, 510 pp., £18.39 (pbk), ISBN 9781032462035. *Teacher Development*, *27*(5), 664–667. https://doi.org/10.1080/13664530.2023.2237484

Zhai, X., Chu, X., Chai, C. S., Jong, M. S. Y., Istenic, A., Spector, M., Liu, J., Yuan, J., & Li, Y. (2021). A review of Artificial Intelligence (AI) in education from 2010 to 2020. *Complexity*, *2021*, 1–18. https://doi.org/10.1155/2021/8812542

6

Artificial-Intelligence-Based Automated Instructor Feedback System for Education and Research

Vinita Yadav and Pradeep Tomar

6.1 Introduction

Consider a real-life scenario where a professor at a college spends weeks examining research papers for a conference, balancing this task with his own research and teaching. Educators are under a lot of pressure, and their students or peers often receive feedback that is either delayed, inconsistent, or lacking in depth evaluation. What if an AI system could pre-evaluate research papers, pointing out key strengths and weaknesses. This is no fiction but the power of an AI-based automated feedback system.

The key to unlock human potential is working smarter with the help of intelligent machines.

Feedback, one of the most powerful means to support learning (Hattie & Timperley 2007), refers to "all post-response information that is provided to a learner to inform the learner about his or her actual state of learning or performance" (Narciss, 2008). In formal education, feedback is often provided by instructors, such as teachers or lecturers, or peer learners, with especially the latter offering high potential for learning (Double et al., 2020).

Feedback – a process where learners make sense of the provided information to reduce the gap between their current and desired performance – is a crucial component of student learning (Watling & Ginsburg, 2018).

For example, feedback to improve students' logical problems can be conveyed by verbally explaining, solution demonstrating, or both. As each student has their own condition and may be differently equipped to access, understand, and use their feedback, both feedback information and communication strategy need to be considered to maximize the benefit of feedback to students and provide a lasting change; hence, the importance of personalized feedback (Kochmar et al., 2020).

 DOI: 10.1201/9781003560814-6

Instructors have a role to effectively formulate and communicate their feedback, while students use the information to update their knowledge and change the corresponding behaviour (e.g., learning strategy, approach to the task, and the use of learning resources) to achieve the desired outcomes(s) (Boud & Molloy, 2013; Forsythe & Johnson, 2017).

Feedback is used for both formative and summative purposes during the learning process. Instructors use formative feedback to provide students with opportunities for continual improvement while using summative feedback to inform students about their performance in the course (Barana et al., 2019; Marriott & Teoh, 2012).

Quality feedback justifies grades, clarifies students' strengths and areas needing improvement, and develops students' learning skills. The limited quality of feedback may demotivate students' understanding and uptake of feedback, harm student–instructor relationships, and thus impact student success (Middleton et al., 2023).

Challenges experienced by instructors when working to provide effective feedback to students include high workload as a result of large classes and institutional constraints, in addition to the online shift due to the COVID-19 pandemic, student action (or inaction), and their own affect and mindset. Large class sizes have also proved a significant barrier to providing effective feedback. While workload and time to provide feedback influenced when and how feedback was provided, student inaction also had a major impact on the feedback cycle, at times completely breaking the cycle. Students' lack of action on feedback resulted in an effective response from the instructors, who felt frustrated due to the lack of student engagement with the feedback, despite the time and effort that the instructor had invested.

Traditional feedback mechanisms often lack effectiveness because they depend on manual processes, which can be time-consuming, inconsistent, and difficult to use. Generic feedback often falls short of expectations in fulfilling individual needs.

As advancements in technology have enabled new ways of learning and changed the dynamics of education (e.g., the transition from traditional paper-and-pencil assessments to digital online assessments), the frequency, delivery format, and timeliness of feedback derived from educational assessments have also changed progressively to meet the needs of students (Jurs & Špehte, 2021). The main goals for using automatic feedback generation are to help students on a specific content/course, support self-regulation, help instructors, and reduce plagiarism behaviour. The purpose of these systems is to analyse large amounts of data, recognize patterns, and produce insights that can improve teaching and research. Digital score reporting has enabled students to receive immediate and personalized feedback from computerized assessments to best inform students in their learning (Bulut et al., 2019). Intelligent tutoring systems (ITSs) can provide granular and specific feedback to students as they complete learning tasks personalized based on their unique interests and proficiency levels (AI, 2017).

6.2 AI in Educational Feedback

Traditional feedback systems are time consuming, with inconsistencies related to subjective interpretation leading to unfair grading. Grant funding organizations in research education receive proposals making it difficult for human reviewers to analyse and produce valuable feedback. New researchers may encounter feedback that is too advanced and unable to meet their personalized needs for improvement.

AI itself refers to the development of computer systems that can perform education-related tasks that require human intelligence, such as grading students' exams, personalizing learning materials, or providing recommendations for assignment tasks based on real-time data analysis (L. Chen et al., 2020). In the education field, AI also plays a role in supporting feedback practices by providing fully automated or semi-automated feedback in various forms such as written feedback (Yu et al., 2019), audio-based feedback (Rodway-Dyer et al., 2011), or video-based feedback (Ketchum et al., 2020). The technology that allows for enhancing the provision of verbal feedback is both manually written by instructors based on the results given by AI (i.e., semi-automated process) or produced by computers (i.e., fully automated process) and non-verbal feedback like computer-generated graphs.

ScholarOne utilizes AI to perform a preliminary screening of research papers, detecting possible problems (e. g., plagiarism, formatting mistakes) prior to their examination by human reviewers. This accelerates the review process and reduces the burden for reviewers. Turnitin Gradescope has the capability to grade assignments for large classes which helps reviewers to concentrate more on personalized and high-value feedback. AI tools such as Publons employ machine learning to evaluate peer reviews and detect trends of bias or inconsistency, assisting journals in upholding elevated standards of fairness and objectivity. ResearchGate utilizes AI to link researchers with comparable interests and offer feedback on one another's work, promoting collaboration and the sharing of knowledge. The emergence of online learning platforms such as Coursera and edX has generated a need for scalable feedback solutions capable of accommodating millions of learners globally. The rapid expansion of scientific research has resulted in an excessive volume of papers and proposals, requiring AI tools to optimize the review and feedback process. By tackling the shortcomings of conventional feedback approaches, AI-driven systems are not only enhancing efficiency and consistency but also allowing for new opportunities for personalized and instantaneous feedback. These innovations are changing the way feedback is provided and applied, ultimately improving results in education and research.

Some examples of applications of AI in providing feedback include the application of machine learning (ML) and natural language processing (NLP) to evaluate the performance of a student in real time to determine which students will likely struggle (Jimenez & Boser, 2021). Educators can also leverage LA, which involves using AI and relevant techniques to deliver real-time personalized feedback to all learners and therefore enhance the quality of their learning experience (Tsai et al., 2021). Such technologies can make it possible for learners to leverage such feedback because of the speed and efficacy of AI applications (Zhang et al., 2019).

In sum, using AI allows educators to provide feedback to a large number of students in a short time frame in contexts such as Massive Open Online Courses (MOOCs), minimal disruptions from time and space barriers, and an ability to process large-scale educational data such as Trends in International Mathematics and Science Study (TIMSS) and Programme for International Student Assessment (PISA) (e.g., Organisation for Economic Co-Operation and Development [OECD], 2019) with educational data mining (EDM) and ML methods. It is important to inform both researchers and practitioners about the present and future of AI applications in feedback practice, given the increasing use of advanced technologies and AI in education world-wide (United Nations Educational, Scientific and Cultural Organization [UNESCO], 2019).

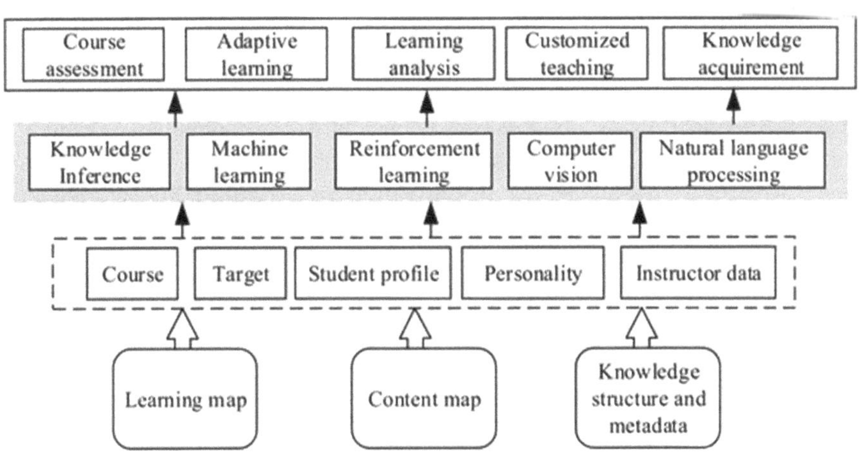

FIGURE 6.1
Technological structure of AI in education. (Lijia Chen et al., 2020.)

The early application of AI for feedback purposes dates back to the 1950s, during which time AI was used for adaptive learning (the self-adaptive keyboard instructor) or computerized assessment. The combination of AI technology and high-quality human instruction allows students to learn more efficiently while at the same time allowing instructors to address issues that can only be identified through results from a high-level data analysis. AI can be used to grade a large number of exams and at the same time identify patterns of student performance with data mining to inform teachers in their feedback provision, such as providing more feedback in the content areas in which the student cohort did not do well. Additionally, a number of testing organisations such as Educational Testing Service (ETS) and Pearson have implemented an automatic essay scoring system (AES) to assess written essays from test takers for a more efficient workflow; the system evaluates the essay based on elements such as grammatical error, writing style, and discourse structure not only to ease the scoring process but also to provide relevant feedback to the test takers for improvements such as their grammar usage, vocabulary diversity, or essay organization. Aside from the educational assessment area, the intelligent tutoring system (ITS) has been used to provide corrective feedback and suggestions to students and human tutors as an enhanced teaching practice. A meta-analysis of 50 controlled evaluations of ITS found that students who received assistance from ITS exhibited greater performance than students from conventional human-only classes (Kulik & Fletcher, 2016).

6.3 The Four Levels of Feedback Framework

Feedback is an essential element in student learning as it helps facilitate student development by stimulating their learning process and optimizing their understanding of class materials for improved performance in the task (Hounsell et al., 2007). In a feedback spiral, students reflect on feedback from their instructors to update their task-related knowledge and behaviour in response to the received feedback; for example, students who receive feedback from their mid-term exam can use it to adjust their learning strategy, such as by investing more time to study the course content they did not do well in to prepare for the final exam (Carless and Boud, 2019). Characteristics of high-quality feedback include thorough coverage, appropriate tone, straightforward language, and transparency in its guidance. Feedback should also be timely and relevant to both the course itself and student circumstances to maximize its actionability, especially in the distance learning context where communication is impeded by the lack of physical proximity.

TABLE 6.1
The Four Levels of Feedback

Feedback Level	The Effect on Students' Level of Change
Task Level	This level concerns how the tasks are performed (e.g., correctly vs. incorrectly)
Process Level	This level concerns the thought process needed to perform the task and its related variant
Self-Regulated Learning Level	This level concerns how students monitor, direct, and regulate their actions towards learning goals
Self Level	This level concerns personal aspects of the students themselves (e.g., goals achieved)

Source: Tarid Wongvorachan et al., 2022.

6.4 AI-Based Automated Feedback Systems

AI-based automated feedback systems are intricate, multi-faceted tools that depend on a blend of technologies and methodologies to evaluate data, create insights, and provide actionable feedback. Below is a detailed examination of the essential components that contribute to the effectiveness of these systems.

6.4.1 Data Collection and Preprocessing

The bedrock of any AI-based feedback system is high-quality data. This element involves acquiring pertinent data and preparing it for analysis. In education, data may originate from student assignments, quizzes, examinations, discussion forums, and learning management systems (LMS). For instance, an AI system could assess essays, coding tasks, or multiple-choice answers. Research data origins encompass research papers, grant applications, peer evaluations, and citation networks. AI systems can also examine datasets, experimental findings, and metadata from scholarly journals.

6.4.2 Data Preprocessing

Cleaning: Eliminating noise, such as irrelevant details or inaccuracies, from the data. For instance, in text-based feedback systems, preprocessing could involve rectifying spelling errors or discarding stop words.

Structuring: Arranging data into a format appropriate for analysis. This could entail transforming unstructured text into structured data using methods such as tokenization or parsing.

Annotation: Tagging data to train supervised machine learning models. For example, essays might be labelled with scores or feedback comments to assist the AI in learning how to assess similar inputs. Example: In an automated essay grading system, raw essays are gathered, cleaned (e. g., correcting typos), and labelled with scores given by human evaluators. These preprocessed data are subsequently utilized to train the AI model.

6.4.3 Machine Learning and Natural Language Processing (NLP) Techniques

AI-based feedback systems are profoundly reliant on ML and NLP to analyse data and produce insights. These techniques allow the system to comprehend context, recognize patterns, and deliver meaningful feedback.

ML-Enabled Enhanced Assessment

Machine learning algorithms thus build using massive databases of previous instructor feedback and grading in order to develop models that have the capacity to analyse submissions by students or research automatically and determine the quality, structure, and coherence of submissions. Such models can identify strengths and weaknesses and suggest areas for improvement.

Rubric-based scoring refers to the process of assessing student work according to predetermined criteria. The majority of rubrics establish specific dimensions or categories, such as clarity, organizational ability, creativity, or guidelines. Often, these dimensions are elaborated upon and accompanied by a scoring scale; for example, 1–5 or "excellent" to "needs improvement", which will characterize levels of performance. This will ensure that the grading is fair, neutral, and transparent; expectations for each part of the work are clearly stated. They determine how well the submission meets the established criteria, such as clarity, organization, and compliance with requirements of the assignment.

Anomaly detection in an automatic feedback generator will be directed towards detecting unusual or not exactly expected patterns within students' submissions. It will answer errors or inconsistencies that may point to issues like plagiarism, failure to follow the guidelines of the assignment, or unexpected responses. Anomaly detection increases the value of the feedback through context-aware suggestions where issues that may be overlooked by normal feedback methods may be highlighted. This adds another layer of intelligence, so the quality and relevance of the overall feedback are enhanced. For example, a drastic shift in the writing style or tone of a student's assignment may alert the anomaly detection system as an indication that it is copied content. Where there are anomalies in the basic structure or logic flow of a submission not commonly seen, this may also trigger a flag to take it under further review.

This comparative analysis is used while generating feedback in comparing the latest submissions made by a student with their previous work to check for progression and identify where the student has to improve. By doing this multiple submission comparison, the system can detect both patterns of growth and future problems that need to be addressed, enabling the delivery of more focused and individualized feedback. This not only allows for more consistency in the feedback but also improvement in writing structure or sense flow can be readily identified and appreciated. Persistent errors here would mean frequent grammatical mistakes or weak arguments, which can be addressed to steer improvements by the student. This allows for steady progress and offers a more lively, contextual experience for feeding back by focusing on the trajectory of the student over time.

The predictive modelling in generating feedback uses historical data and machine learning algorithms to predict possible outcomes or behaviours based on the established patterns that have been prevalent in previous submissions. This allows the system to predict areas in which a student might face challenges or excel at, making it proactive enough in giving targeted feedback in advance. On another note, if there is a specific kind of assignment that a student finds too frustrating most of the time, or in which they face intense learning issues with a certain subject or idea, the model would expect those future frustrations and, consequently, offer alternative resources or options to mitigate that problem before it worsened. Predictive models can also identify trends wherein improvements are expected based on patterns in the work output and guide the channelling of feedback. They provide individualized feedback to the student on their learning journey and improve the educational experience by resolving problems before they arise.

Natural Language Processing (NLP)

AI-driven feedback systems use NLP techniques to analyse and understand the content of student assignments, research papers, or other educational and research materials. This includes extracting semantic meaning, identifying key concepts, and detecting common patterns and errors.

Semantic Analysis: This is the interpretation of the deep meaning, concepts, and ideas of the text, rather than words and phrases in the outer layer (Crossley et al., 2017). Semantic analysis involves the analysis of a grammatical structure of sentences with regard to an order or position of words, phrases, and clauses in them through which it identifies the relationships between independent terms within a specific context. This is a very critical task carried out by NLP systems. It is also the core of many machine learning applications in use, whether it is an online search, a chatting bot, or other text-analysing software.

Syntactic Analysis or Parsing or in the Phase of NLP: This phase is aimed at drawing an exact meaning from the text. Syntax analysis checks the text for meaningfulness compared to the rules of formal grammar. Syntactic Parsing: This is the process of analysis of the grammatical structure and relationship between the various components of the text.

The structured, collective, and coherent groups of sentences in NLP are described as discourse. In the field of NLP, the inherent relationships between words can simplify the training process of NLP models and make their behaviour more predictable—though not necessarily their outcomes. This is particularly relevant when analysing the overall flow, structure, and coherence of written text (Chakraborty et al., 2018).

To categorize the types of NLP errors, techniques are utilized such as Parse Errors Syntax or grammatical error detection which target the wrong word ordering and incorrect application of punctuation/full stops. Semantic errors are understanding inconsistencies such as ambiguous phrases or wrong word selection. Error identification includes repeated grammatical, spelling, and punctuation mistakes (Crossley et al., 2017).

The system achieves a deep understanding of the content and quality of the submitted materials by utilizing NLP models such as transformer-based language models (e.g., BERT, GPT). Example: Turnitin's AI-driven feedback tool employs NLP to evaluate student essays, identify plagiarism, and provide recommendations for enhancing grammar, style, and structure.

6.4.4 Feedback Generation and Delivery

After the data have been examined, the system produces feedback and conveys it to the user in a straightforward and actionable manner.

Feedback Generation

Rule-Based Systems: These systems utilize pre-established rules to formulate feedback. For instance, a coding feedback tool may assess syntax errors and suggest corrections based on programming rules.

AI models create feedback based on patterns acquired from data. For example, an essay evaluation system might offer insights on coherence, argumentative strength, and grammar.

Hybrid Systems: Merging rule-based and model-based methods to deliver more all-encompassing feedback. For instance, a system could use rules to identify fundamental errors and a model to offer advanced suggestions.

6.4.5 Feedback Delivery

Text-Based Feedback: Written notes or recommendations, such as grammar adjustments or improvement advice.

Visual Feedback: Graphs, diagrams, or heatmaps to emphasize areas needing enhancement. For example, a writing feedback tool might implement a heatmap to showcase frequently used words or phrases.

Interactive Feedback: Immediate suggestions while the user is working. For instance, a coding platform could underline mistakes and propose corrections as the student types.

Multimodal Feedback: Merging text, visuals, and audio to accommodate various learning preferences. For example, a language learning app might present audio pronunciation input alongside written corrections. Example: Grammarly provides instantaneous, text-based feedback on

TABLE 6.2

NLP Techniques—ML Algorithms

NLP Technique	ML Algorithm
Methodology	
Feature extraction	BERT, LSA, LDA,CNN, LSTM,
Feature selection	Naive Bayes, Chi Square, SVM, Random Forest
Topic modelling	LSA, NMF, LDA,PLSA
Text evaluation	
Text summarization	LSTN-NN,CNN, Bi-GRU, Text Rank, LSA, BERT, LSTM-CNN
Document categorization	LSTM, CNN, BLSTM-C
Entity extraction	CNN
Knowledge graphs	Multidimensional knowledge graphs
Sentiment annotation	CNN, LSTM

Source: Thanveer Shaik et al. 2022.

grammar, tone, and style while users write. It also offers visual cues (e. g., underlines) to signify issues.

Integration with Existing Systems: For AI-driven feedback systems to function effectively, they must integrate smoothly with current platforms and workflows. AI feedback tools can be incorporated into LMS platforms like Canvas or Moodle to deliver automated grading and feedback for assignments and quizzes. AI tools can be included in academic publishing platforms to aid in paper review, plagiarism detection, and citation analysis.

Example: Elsevier's AI-enabled tools assist researchers in locating relevant literature and improving the quality of their manuscripts before submission.

6.5 AI Technologies in Educational and Research Feedback: Applicative Field

AI is an umbrella term covering a wide area of machine capability, from basic problem solving such as rule-based message delivery to advanced decision-making including multi-class machine learning-based classification. The most relevant AI technologies for feedback are situated in NLP, EDM, and learning analytics (LA) (Nguyen et al., 2020).

Table 6.3 presents the application of AI technologies in these three research fields in terms of their definition and capability.

The application of AI in NLP focuses on manipulating unstructured textual data for understanding, interpreting, and potentially generating relevant textual output. Textual data used in NLP can come in any shape ranging from simple words or sentences such as student-authored course reviews to

TABLE 6.3

The capability of AI in Three Fields

Research Field	Focus of the Application of AI Technology	Capability
NLP	The understanding and manipulation of textual data	Convert textual data for translation or pattern extraction
EDM	Knowledge extraction from educational databases	Automatically extracting information using machine learning to discover insights
LA	The leveraging of student activity data for classroom optimization. It aims at discovering concealed patterns within the raw gathered data from academic educational settings	Process learning activity data to support human judgement in classrooms. Learning analytics converts unstructured data into particular knowledge

Source: Tarid Wongvorachan et al., 2022.

complex essays with various structures and writing styles from GRE/GMAT examinees. Algorithms used in NLP can automatically convert the data into understandable formats and extract non-trivial information from it by analysing elements such as syntax, semantics, morphology, or even the basic frequency of words. Some applications of NLP in education include text summarization to extract essential elements from unstructured documents (e.g., theses, essays, or reports), machine translation to bypass or mitigate language barriers, and sentiment analysis to gain insights into public opinion.

The application of AI in EDM focuses on using ML techniques such as clustering or classification of educational databases for knowledge discovery. Data used in EDM can be both numerical and textual in nature and can take various forms, such as student performance as indicated by their GPA, their history of grade repetition as indicated by self-reported binary indicator (i.e., yes vs. no), or even the educational level of their parents; such data can come from large-scale sources such as student records or educational surveys (e.g., PISA, TIMSS) . Some applications of EDM include the prediction of student performance from school databases to identify potential low-achieving students or the development of adaptive learning systems with student log data to provide personalized lessons. Such insights can support stakeholders in the educational context, such as principals, teachers, or even parents, to make informed decisions on matters such as curriculum design or school development.

The application of AI in LA focuses on collecting and leveraging accumulated data on student learning processes and activities for classroom optimization. Similar to EDM, the research focus of LA lies in the translation of data-driven insights into practical recommendations to guide the process of planning, decision-making, and intervention. Like EDM, LA can also process numerical and textual data. However, the difference between the two fields is that LA operates from a holistic framework that considers student data as a whole while performing descriptive and diagnostic analyses with an emphasis on the process of teaching and learning. On the other hand, EDM focuses more on knowledge discovery from various analysis techniques. Learning analytics can help in identifying the most suitable educational conceptual framework for a designated project or course. Universities can uncover and develop effective strategies to improve specific practices and curriculum goals by connecting resources utilized by students and staff with particular educational objectives. Educators can utilize learning analytics to pinpoint the courses that are closely associated with a student's summative assessment and those linked to significant variations in performance and curriculum attrition. Educators can swiftly recognize which assessments are effective and require revision. Learning analytics can support students in achieving success by providing them with relevant statistical data and promptly informing them if they are in danger of losing a program (Monika Hooda et Al. 2022).

EDM researchers focus more on the automated discovery aspect, such as predictive or descriptive analytics and machine learning models. In contrast, LA researchers focus on leveraging data to inform human judgement, such as investigating data patterns, resource allocation, and their effect on learning and teaching practice.

6.6 AI-Automated Feedback System: Features

In an ideal AI-based automatic feedback system, its components usually contain the following:

- The assessment can be made with detail on content, structure, and quality in a given student assignment or research paper that an AI model trains to undertake with specific scores according to attached rubrics for automatic assignment and grading.
- The system grades submissions to report areas of strong strengths, specific weaknesses, and particular areas for further work by pointing out common grammatical mistakes, logical inconsistency, missing citations, or inadequately designed experiments produced with the aim of identifying areas of strength, weakness, and potential for improvement.
- The system performs the analysis of work done, it would allow personal feedback to be given to the student or researcher, providing more particulars of comments and suggestions about how to deal with the problematic issues, thus improving the quality of the work.

- The feedback system will be able to monitor the learner's progress for a particular student or researcher over time to enable the instructors to identify emerging trends on the part of the learning of the students and thus grounds for customized interventions or adjustments in their teaching or research strategies.

6.7 Challenges and Implementation Considerations

AI-based automated feedback systems require addressing several essential considerations during their implementation phase.

6.7.1 Bias in AI Algorithms

A significant ethical concern in AI-driven evaluations is the risk of bias. The impartiality of AI systems is directly linked to the quality of the data they are trained on. If the training data contains biases – whether concerning gender, race, socioeconomic background, or disability – the AI may reproduce or even heighten these biases.

For instance, a case study (Ehsan Latif et al., 2025) demonstrates that AI models trained on imbalanced gender data can show biases, resulting in differences in scoring between male and female students. This is an issue because biased AI systems can perpetuate societal stereotypes and inequalities, thus impacting students' educational and career prospects. This is problematic because biased AI systems can reinforce societal stereotypes and inequalities, thereby affecting students' academic and career opportunities.

Ethical considerations such as ensuring fairness, transparency, and the use of inclusive training data are crucial to prevent discrimination and promote equal opportunities. Continuous monitoring and improvement of AI systems are essential to maintain trust and fairness in educational assessments.

6.7.2 Privacy and Data Security

AI-driven assessments often involve extensive data collection, including students' performance metrics and personal information. This raises significant privacy and data security concerns as unauthorized access or misuse of these data can lead to breaches of confidentiality and privacy. For example, in 2020, a major online learning platform faced a data breach that exposed the personal information of thousands of its students. Institutions must adhere to strict data protection regulations, such as GDPR or FERPA, to ensure that robust security measures are in place to safeguard student data.

Clear policies on data usage and consent should be established to maintain transparency.

A scientist utilizing an AI application to examine confidential health data unintentionally revealed patient details because of a vulnerability in the system. This incident breached HIPAA (Health Insurance Portability and Accountability Act) guidelines and undermined confidence in the research organization.

Challenge: Research information, particularly in areas such as healthcare or social sciences, frequently contains sensitive details that require careful handling. AI technologies must guarantee strong data encryption and access controls.

6.7.3 Transparency and Accountability

AI systems can often operate as "black boxes" in which the decision-making process is opaque. This lack of transparency can undermine trust and make it difficult for educators and students to understand how assessments are determined. Developers should provide clear explanations of how AI systems make decisions and offer insights into the data and algorithms used.

A student got a poor score on an essay from an AI evaluation tool but could not comprehend the reason. The absence of clarity in the feedback process made the student feel exasperated and unable to enhance their work.

Challenge: AI systems need to deliver transparent and understandable feedback, allowing users to grasp the reasoning behind their judgements. This is particularly critical in high-stakes situations such as grading or admissions. Accountability measures, such as regular reviews and third-party evaluations, can also help ensure that AI-driven assessments are fair and accurate.

6.7.4 Accuracy and Reliability

While AI can enhance the efficiency of assessments, it is vital to ensure that the systems themselves are accurate and reliable, because errors in AI-driven assessments can lead to incorrect grading or feedback that can impact students' educational outcomes. The continuous testing and validation of these AI systems are necessary in order to maintain optimal standards of accuracy. Feedback mechanisms must be in place to address and rectify any discrepancies in assessment results.

In 2020, the United Kingdom's A-level examination outcomes were established by a computer algorithm following the cancellation of in-person exams due to COVID-19. The algorithm reduced almost 40% of forecasted grades, with a disproportionate impact on students from less privileged backgrounds. This resulted in significant public anger and a change in the decision, underscoring the dangers of depending on AI for critical decisions.

Challenge: AI systems could find it difficult to grasp context, nuance, or creativity, resulting in inaccurate assessments. For example, an AI grading program might incorrectly perceive a student's unusual essay format as sub-standard writing.

6.7.5 Equity of Access

AI-driven assessments and feedback should be accessible to all students, including those with disabilities or limited access to technology. Equity of access is a fundamental ethical consideration, as it ensures that all students have an equal opportunity to benefit from assessment tools. Online assessment solutions and assessment development should incorporate features that accommodate diverse learning needs and technological access. This includes providing alternative formats and ensuring that platforms are usable by individuals with disabilities.

Researchers in developing nations frequently do not have access to sophisticated AI tools because of expensive prices or poor internet access. This worsens the already present disparities within the global research community.

Challenge: AI tools should be created with inclusivity as a priority, guaranteeing that researchers from diverse backgrounds can take advantage of their features.

Ethical concerns must be addressed thoroughly, emphasizing fairness, transparency, and accountability, and minimizing biases within AI models. Some educational institutions have encountered backlash for employing AI systems to oversee student conduct, including observing eye movements while students take online tests. Detractors claim that this infringes on students' privacy and fosters a climate of surveillance.

Challenge: AI systems should reconcile the necessity for accountability with respect for students' independence and privacy. This necessitates gaining informed consent and making certain that AI tools are utilized ethically.

Incorporating instructors' domain expertise and feedback practices into system design and model training is essential to ensure that automated feedback remains accurate and relevant.

Balancing Automation and Human Oversight: Although AI can handle numerous feedback tasks, human oversight and the option to intervene or offer supplementary feedback are crucial for maintaining quality and fostering trust in the system.

Ensuring Transparency and Explainability: The system should be designed to offer clear explanations for its feedback, enabling users to understand the reasoning behind its suggestions and recommendations.

6.8 Case Study

AI-based automated feedback systems have a wide range of applications in education and research, including the following.

6.8.1 Automated Essay Grading in Education

Example: Turnitin's Gradescope
Gradescope is a tool powered by AI that streamlines the grading process for essays, tests, and assignments. It has been extensively utilized by educational institutions to conserve time and deliver uniform feedback. A professor at a university utilized Gradescope to assess 200 essays in less than one hour, in contrast to several days required for manual grading. Students obtained thorough feedback on their essays, which included recommendations for enhancing grammar, organization, and the strength of arguments.

Challenges: The platform initially faced difficulties in assessing creative or non-traditional writing styles, resulting in student grievances. Concerns emerged regarding the opacity in the grading process and how scores were established.

Considerations: AI grading solutions need to be crafted to accommodate varied writing styles and offer transparent rationales for their assessments. Ongoing enhancement and human oversight are vital to guarantee equity and precision.

6.8.2 AI-Enhanced Peer Review in Research (Bauer, 2023)

Example: Publons
Publons employs AI technology to facilitate peer review, aiding journals and conferences in streamlining the review process more effectively. A prominent academic journal shortened its peer review duration from six months to three months by utilizing Publons' AI tools for pre-screening submissions. Reviewers received AI-generated recommendations for focal points, enhancing the quality and consistency of the feedback provided.

Challenges: Some reviewers voiced concerns regarding the possibility of AI substituting human judgement, resulting in resistance to its implementation. The system sometimes identified high-quality papers as low-priority due to its dependence on established criteria.

Considerations: AI should augment, not replace, human expertise in the peer review process. Transparency and clarity are essential for building trust and acceptance among the research community.

6.8.3 Personalized Learning Feedback in K-12 Education

Example: DreamBox
DreamBox is a customizable learning platform that employs AI to deliver tailored feedback and suggestions to K-12 learners.

A school district in California noted a 20% increase in maths scores following the adoption of DreamBox. Students received instantaneous feedback on their performance, assisting them in recognizing and addressing areas of weakness.

Challenges: Some educators experienced concerns that the system diminished their role in the classroom, raising worries about job security. The platform necessitated substantial technical infrastructure, which posed a challenge for underprivileged schools.

Considerations: AI tools need to be created to support, not substitute, teachers. Efforts should be made to guarantee fair access to AI-enhanced educational tools.

6.8.4 Real-Time Writing Feedback for Researchers

Example: Grammarly
Grammarly is an AI-driven writing assistant that offers immediate feedback on grammar, style, and tone. A PhD candidate utilized Grammarly to enhance their dissertation, leading to a clearer and more refined final draft. Researchers indicated that the tool assisted them in enhancing the readability and impact of their publications.

Challenges: Some users became excessively dependent on the tool, which could impede their ability to cultivate independent writing skills. The system sometimes delivered inaccurate suggestions, especially for technical or specific domain writing.

Considerations: AI writing tools ought to be utilized as support, rather than crutches, to encourage independent learning and skill enhancement. Customization for specific fields is essential to boost accuracy and relevance.

6.9 AI-Enabled Feedback System: Challenges

As AI-driven automated feedback systems progress, numerous prospective trends and challenges are becoming apparent, including the integration of multimodal feedback. Incorporating feedback across various formats – such as textual, auditory, and visual – can yield a more thorough and engaging experience for users. These challenges influence the dependability, accessibility, and efficacy of feedback systems designed to augment human input in educational environments. The development of AI models capable of delivering clear, interpretable rationales for their feedback will be essential for cultivating trust and promoting meaningful learning experiences.

AI systems are substantially dependent on extensive quantities of high-quality, annotated data to facilitate learning and feedback generation. In numerous instances, educational datasets may exhibit a lack of diversity, be outdated, or not exist in adequate volumes, thereby impeding precise model training.

Should the training data exhibit biases (e.g., cultural or socioeconomic biases), the system may inadvertently produce feedback that is biased, thereby impacting the fairness and inclusivity of the feedback (Jimenez & Boser, 2021). Algorithms may unintentionally prioritize specific demographics due to the inherent biases in the training data. To ensure equitable feedback for varied student populations, it is imperative to implement rigorous monitoring and iterative refinement of algorithms (Middleton et al., 2023).

Feedback systems frequently gather sensitive student information, rendering them vulnerable to privacy concerns. Achieving compliance with privacy regulations (such as GDPR) while simultaneously delivering personalized feedback represents a complex equilibrium (United Nations Educational, Scientific and Cultural Organization [UNESCO], 2019).

The feedback generated by AI must correspond with the learning objectives and pedagogical strategies employed. Feedback that is misaligned may lead to confusion among students and diminish instructional efficacy. Although automated feedback is efficient, students typically gain from the personalized approach and empathy that characterize human feedback. Developing a system that harmonizes automation with personalized, human-like attributes poses a considerable challenge (Boud & Molloy, 2013).

6.10 Conclusion

This chapter focuses on examining current AI applications for feedback in educational and research settings to pinpoint areas where further research can advance technology-driven feedback practices. Advanced NLP and ML technologies allow feedback systems to deliver tailored, immediate, and actionable insights on a large scale, thereby enhancing learning results, boosting instructor efficiency, and streamlining research processes. As the field of AI in education evolves, understanding essential implementation factors and embracing upcoming trends will be pivotal for the successful integration of these transformative technologies in educational and research domains.

Initially, AI in education (AIED) was limited to basic computing systems and later evolved into web-based and online platforms. These advancements have since led to embedded systems, such as cobots (collaborative robots) and humanoid robots, which now function alongside or independently of teachers. Chatbots also offer instructional support, performing tasks typically managed by instructors, thereby increasing instructional quality and teacher productivity. For students, AI has enabled the personalization of learning materials to suit diverse needs and learning abilities, providing a richer educational experience.

AI-driven feedback systems, in particular, could harness the capabilities of digital learning platforms to improve the quality of feedback and expedite the feedback process. Real-time, interactive feedback systems embedded in online assessments could foster dynamic exchanges between students and virtual tutors, increasing engagement and motivation.

In essence, AI in feedback can exemplify best practices in effective feedback design, content quality, and the inclusion of supportive materials for a comprehensive educational experience.

References

Bauer, Elisabeth, Markus Berndt, Martin R. Fischer, and Frank Fischer, Using natural language processing to support peer-feedback in the age of artificial intelligence: A cross-disciplinary framework and a research agenda, *British Journal of Education Technology*, 54, 1224-1245, 2023.

Boud, David and Molloy, Elizabeth, Rethinking models of feedback for learning: the challenge of designAssessment and Evaluation in Higher Education, 38(6), 698-712, 2013 DOI: 10.1080/02602938.2012.691462

Carless, David, and David Boud, The development of student feedback literacy: Enabling uptake of feedback, *Assessment & Evaluation in Higher Education*, 2018.

Cavalcanti, Arthur Barbosa, Ruan Carvalho, Fred Freitas, Yi-Shan Tsai, Dragan Gašević, and Rafael Ferreira Mello, Automatic feedback in online learning environments: A systematic literature review, *Computers and Education: Artificial Intelligence*, 2, 100027, 2021.

Chakraborty, Koyel, Rajib Bag, and Siddhartha Bhattacharyya, Relook into sentiment analysis performed on Indian Languages using Deep Learning, In *The Proceedings of 2018 Fourth International Conference on Research in Computational Intelligence and Communication Networks (ICRCICN)*, IEEE, 208-213, 2018, doi: 10.1109/ICRCICN.2018.8718709.

Chen, Lijia, Pingping Chen, and Zhijian Lin, Artificial Intelligence in education: A review, *IEEE Access*, 8, 2020. DOI 10.1109/ACCESS.2020.2988510

Crossley, Scott A., Kristopher Kyle, and Danielle S. McNamara, Sentiment Analysis and Social Cognition Engine (SEANCE): An automatic tool for sentiment, social cognition, and social-order analysis, *Behavior Research Methods*, 49, 803-821, 2017.

Dai, Wei, Yi-Shan Tsai, Jionghao Lin, Ahmad Aldino, Hua Jin, Tongguang Li, Dragan Gašević, and Guanliang Chen, Assessing the proficiency of large language models in automatic feedback generation: An evaluation study, *Computers and Education: Artificial Intelligence*, 7, 100299, 2024.

Deng, Ruiqi, and Pierre Benckendorff, *Technolog-Enabled Learning,* Handbook of e-Tourism, 2020, doi:10.1007/978-3-030-05324-6_129-1

Double Kit S., Joshua A. McGrane, Therese N. Hopfenbeck, The Impact of Peer Assessment on Academic Performance: A Meta-analysis of Control Group Studies. *Educational Psychology Review* (2020) 32, 481–509. https://doi.org/10.1007/s10648-019-09510-3

Forsythe Alex and Johnson Sophie, Thanks, but no thanks for the feedback, Assessment & Evaluation in Higher Education, 42(6), 1-10. 2017. DOI:10.1080/02602938.2016.1202190

Foy, Pierre, Bethany Fishbein, Matthias von, and Davier Liqun Yin, Implementing the TIMSS 2019 scaling methodology, *Methods and procedures: TIMSS 2019 technical report 1.*

Jimenez, Laura, and Ulrich Boser, *Future of testing in education: Artificial Intelligence,* Center for American Progress, 2021.

Hattie, J., and Timperley, H., The power of feedback. Review of educational research, *Sage Journals*, 77(1), 2007. doi: 10.3102/003465430298487

Holmes, W., M. Bialik, and C. Fadel, Artificial intelligence in education: Promises and implications for teaching and learning, *The Center for Curriculum Redesign*, 2019. ISBN: 1794293701.

Holmes, W., K. Porayska-Pomsta, K. Holstein, E. Sutherland, T. Baker, and S. Buckingham Shum, Ethics of ai in education: Towards a community-wide framework, *International Journal of Artificial Intelligence in Education*, 32, 504-526, 2021.

Hooda, Monika, Chhavi Rana, Omdev Dahiya, and Jayashree Premkumar Shet, Integrating LA and EDM for improving students success in higher education using FCN Algorithm, *Hindawi Mathematical Problems in Engineering*, 2022. doi: 10.1155/2022/7690103

Hounsell Dai, Velda McCune, Jenny Hounsell and Judith Litjens, The quality of guidance and feedback to students, *Higher Education Research & Development*, 27, 55-67, 2007.

Jurs Pavels, Špehte Elita, The Role of Feedback in the Distance Learning Process, Journal of Teacher Education for Sustainability, 23, 91-105, 2021.

Ketchum Cheri, Daria S. LaFave, Chelsey Yeats, Elaine Phompheng, and James H. Hardy, Video-Based Feedback on Student Work: An Investigation into the Instructor Experience, Workload, and Student Evaluations, *Online Learning*, 24(3), 85-105, 2020.

Khosravi, Hassan, Simon Buckingham Shum, Guanliang Chen, Cristina Conati, Yi-Shan Tsai, Judy Kay, Simon Knight, Roberto Martinez-Maldonado, Shazia Sadiq, and Dragan Gašević, Explainable Artificial Intelligence in education, *Computers and Education: Artificial Intelligence*, 3, 1000742022, 2022.

Kochmar Ekaterina, Dung Do Vu, Robert Belfer, Varun Gupta, Iulian Vlad Serban, Joelle Pineau, Automated Personalized Feedback Improves Learning Gains in an Intelligent Tutoring System, *Artificial Intelligence in Education*. 12164, 140–146, 2020. doi: 10.1007/978-3-030-52240-7_26

Kulik, James A., Fletcher, J. D, Effectiveness of intelligent tutoring systems: A meta-analytic review. *Review of Educational Research*, 86(1), 42–78. https://doi.org/10.3102/0034654315581420

Latif, Ehsan, Xiaoming Zhai, and Lei Liu, AI gender bias, disparities, and fairness: Does training data matter?, arXiv:2312.10833v4 [cs.CY], 2025.

Lin, Chien-Chang, Anna Y. Q. Huang, and Owen H. T. Lu, Artificial intelligence in intelligent tutoring systems toward sustainable education: A systematic review, *Smart Learning Environment*, 10, Article number: 41, 2023.

Mathew, Leeja, and V. R. Bindu, A review of natural language processing techniques for sentiment analysis using pre-trained models, In *The proceedings of 2020 Fourth International Conference on Computing Methodologies and Communication (ICCMC)*, IEEE, 2020.

Meyer, Jennifer, Thorben Jansen, Ronja Schiller, Lucas W. Liebenow, Marlene Steinbach, Andrea Horbach, and Johanna Fleckenstein, Using LLMs to bring evidence-based feedback into the classroom: AI-generated feedback increases secondary students' text revision, motivation, and positive emotions, *Computers and Education: Artificial Intelligence*, 6, 2024, 100199. doi: 10.1016/j.caeai.2023.100199

Morales-Chan, Miguel, Hector R. Amado-Salvatierra, José Amelio Medina, Roberto Barchino, Rocael Hernández-Rizzardini, and António Moreira Teixeira, Personalized feedback in massive open online courses: Harnessing the power of LangChain and OpenAI API, *Electronics*, 13(10), Article 1960, 2024. doi: 10.3390/electronics13101960

Nazaretsky, Tanya, Paola Mejia-Domenzain, Vinitra Swamy, Jibril Frej, and Tanja Käser, AI or human? Evaluating student feedback perceptions in Higher Education, *Accepted to Nineteenth European Conference on Technology En- hanced Learning (ECTEL 2024).*

Narciss, S., Feedback strategies for interactive learning tasks. In *Handbook of research on educational communications and technology*, 125-144 Edition: 3rd ed Chapter: 11, Lawrence Erlbaum Associates, 2008.

Paris, Brit M., Instructors' perspectives of challenges and barriers to providing effective feedback, *Teaching & Learning Inquiry*, 10, 2022. doi: 10.20343/teachlearninqu.10.3

Rodway-Dyer Sue, Jasper Knight, Elizabeth Dunne, A Case Study on Audio Feedback with Geography Undergraduates, *Journal of Geography in Higher Education*, 35(2), 217-231, 2011.

Shi, Huawei, and Vahid Aryadoust, *A systematic review of AI-based automated written feedback research*, Cambridge University Press, 2024

Tao, Thanveer Shaik Xiaouhi, Yan Li, Christopher Dann, Jacqui Mcdonald, Petrae Redmond, and Linda Galligan. Review of the trends and challenges in adopting natural language processing methods for education feedback analysis, *IEEE Access*, 10, 56720–56739, 2022.

Tsai, Yi-Shan, Vitomir Kovanović, and Dragan Gašević, *Connecting the dots: An exploratory study on learning analytics adoption factors, experience, and priorities*, The Internet and Higher Education, 50, Article number 100794, 2017.

Wang, Shan, Fang Wang, Zhen Zhu, Jingxuan Wang, Tam Tran, and Zhao Du, Artificial intelligence in education: A systematic literature review, *Expert Systems With Applications*, 252, Part A, Article number 124167, 2024.

Watling, Christopher J., and Shiphra Ginsburg, *Assessment, feedback and the alchemy of learning, Medical Education*, 53(1), 76-85, 2018. doi: 10.1111/medu.13645. Epub 2018 Aug 2.

Wongvorachan, Tarid, Ka Wing Lai, Okan Bulut, Yi-Shan Tsai and Guanliang Chen, Artificial Intelligence: Transforming the future of feedback in education, *Journal of Applied Testing Technology*, 23(1), 2022.

Xu, Wenzhong, Jun Meng, S. Kanaga Suba Raja, M. Padma Priya, and M. Kiruthiga Devi, Artificial intelligence in constructing personalised and accurate feedback systems for students, *International Journal of Modeling, Simulation, and Scientific Computing*, 14(1), Article 2341001, 2023. https://doi.org/10.1142/S1793962323410015

7

Enhancing Teacher Preparation for Integrating Computational Thinking Through Unplugged Activities: Utilizing Learning Styles and MBTI Personality Types

Rajesh Kumar Modi, Ayushman Pranav, Umesh Gupta and Ankit Dubey

7.1 Introduction

In today's digital world, experts have written articles around the problem of embedding computers or computational thinking in students. Indeed, fostering such skills at an early stage will help children face future technological obstacles. Since computational thinking includes the intellectual processes of self-organization, recognizing patterns, and working through problems, in order to prepare students for this, they need to be taught effective pedagogical approaches that help embed computational thinking in the context of the 21st century [1]. The skills needed to function in society are quite often computer-based and involve proficiency with technology. Bringing student involvement into play through connected activities makes general classroom activities much more interesting and relevant. To develop computational thinking abilities, it is effective to utilize certain "unplugged activities" that do not involve the student interacting with computers. These exercises allow computational thinking to be developed in classrooms lacking technology resources. There are too many applications, therefore, minimal screen time is certainly advantageous. Training teachers on how to administer these activities is important concerning the tasks' goals. Teachers have different styles of learning (visual, auditory, kinesthetic, reading/writing) and personality types (according to MBTI) which drives the students' interactions in the teaching activities. Providing training that fits these individual features would make it easier for teachers to explain concepts in computational thinking.

As can be seen in Figure 7.1, the flowchart provides a step-by-step outline of how teachers can be assisted in integrating computational thinking using

DOI: 10.1201/9781003560814-7

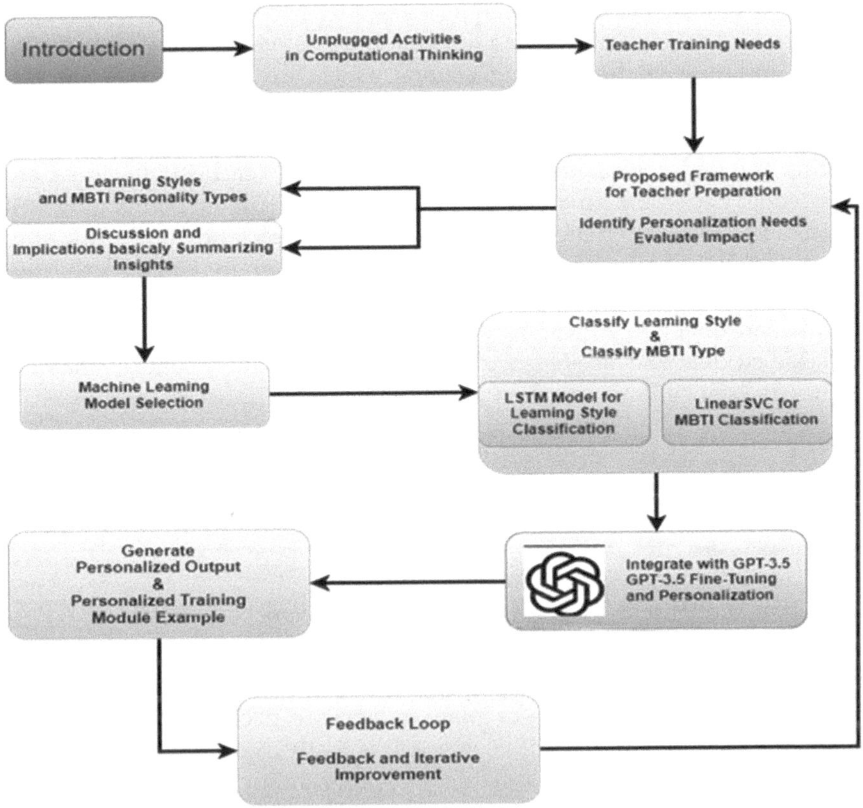

FIGURE 7.1
Enhancing teacher preparation for integrating computational thinking through unplugged activities.

unplugged activities. The process has eight components and starts with contextualizing the Introduction, after which the focus is on the importance of looking for teacher training. The conceptualized framework comprises steps for the determination of the personalization requirements according to the teachers' learning styles and their MBTI types [2],[3]. To achieve this, LSTM and LinearSVC were employed to classify the aspects. This classification is used in the fine-tuning of GPT-3.5, which aims at producing individualized training guidance to appeal to teacher participation and instructional effectiveness. The iterative process also allows the framework to be continuously improved to deliver meaningful and flexible learning experiences to teachers as well as, finally, their learners.

This chapter outlines the very framework which relies upon machine learning and artificial intelligence (AI) to undertake teacher preparation. An LSTM model assigns classes of teachers according to their learning styles, whereas a LinearSVC model assigns classes by MBTI types of personality. Teachers also receive automated invites with tailored, AI-generated summaries outlining skill-enhancing actions which are sure to boost engagement. The use of these computational tools helps in aiding in the execution of unplugged activities in a customized fashion [4]. Employing personalized training methodologies, this approach affords all teachers the tools and knowledge necessary to make a difference and provide engaging and effective learning experiences. In this chapter, we check how the application of learning styles and personality types during teacher preparation can foster computational thinking integration, emphasizing how technology can enable diverse and effective education of this kind. However, the deployment of AI-based models also faces significant challenges, including data availability, teacher resistance, and resource constraints, which must be addressed to ensure successful implementation.

7.2 Understanding Computational Thinking

Computational thinking as a problem-solving strategy consists of decomposing problems, looking for patterns, generalization of principles, and devising detailed procedures or plans which are necessary skills for any person interacting with the digital space. It is firstly a computer science skill, but this involves more than just code and programming, it enables individuals to think reasonably and analytically in multiple situations beyond just programming [5].

7.2.1 Core Concepts of Computational Thinking

Many concepts that are considered core to computational thinking are defined below, each of them plays a distinct role in the process of formulating and resolving problems:

- Decomposition: Breaking down complex problems into smaller, manageable parts, making them easier to solve.
- Pattern Recognition: Identifying similarities or patterns in problems, which can simplify problem-solving and reveal efficient approaches.
- Abstraction: Focusing on the relevant information and ignoring extraneous details, allowing one to generalize and apply solutions to broader contexts.
- Algorithm Design: Creating step-by-step solutions or rules to solve problems, which is the basis of developing programs and automated processes.

Thus, these concepts enable students as well as professionals to address challenges in a stepwise manner beginning from the definition of the problem to the final stage of implementing a solution to the problem [6].

7.2.2 Relevance in the Modern Digital Landscape

In the modern world, which is characterized by technology, it makes sense to say that computational thinking is one of the important skills which students will use in their future careers requiring them to be digitally literate, flexible, and able to think critically. Other than applied sciences or STEM domains, computational thinking is also useful in sectors such as healthcare, business, or arts, as it improves data comprehension, process optimization, and solution generation [7]. Digitalization not only seeks to co-opt every domain of existence but also enables students to comprehensively think and work in a complicated digital world. In the future, computational thinking will be among the basic skills every individual should possess.

7.3 Unplugged Activities: A Teaching Approach

Unplugged activities provide a particular way of teaching computational thinking without the aid of technology. These activities include interactive exercises, games, and physical exercises that help in demonstrating basic computing concepts in an enjoyable and practical manner. To help in the visualization of abstract ideas, unplugged activities ensure that students learn computational thinking in relation to the real world and without devices. This is a useful method when classrooms do not have many technological facilities or where there is sufficient emphasis placed on low screen time [8].

Examples of unplugged activities include:

- Sorting Networks: This activity uses ropes, cones, or markers to physically illustrate the steps of a sorting algorithm. Through a network of steps, students do not just learn to look and compare elements in relation to each other but also touch on how the same relates in the case of being sorted [9].
- Binary Representation: In this case, students comprehend binary numbers and how data can be represented through its basic principles using physical things such as cards or colored blocks. This practical approach makes the student understand binary codes by making students understand coding rather than simply writing it down [10].
- Mazes as a Problem-Solving Tool: In using a floor or board, students create ways to move through the maze, giving them the ability to use

progressively sequential commands. This activity encourages students to employ algorithmic thinking, moving a task through a number of steps, similar to the logic used in computer programming.

These unplugged activities transform abstract notions of computing into practice, and as such, students can gain a greater appreciation of the concepts being taught [9].

7.3.1 Benefits of Unplugged Activities in Teaching Computational Thinking

Unplugged activities offer a range of educational benefits, particularly teaching computational thinking in diverse and inclusive settings. Some key advantages include:

- Enhanced Accessibility: Since unplugged activities do not require digital devices, they can be readily implemented in a variety of settings, including schools with limited access to technology. This unlocks the doors to many people when it comes to education for computational thinking.
- Increased Student Engagement: The hands-on and interactive nature of unplugged activities can emotionally engage students and families. A great deal of enjoyment is derived from playing sport or any active games which signify the construction aspects of the computing concepts for the younger audience [11].
- Reduced Screen Time: There is a strong counterbalance with respect to digital technologies in the context of these unplugged activities. With rising concerns about screen time, these unplugged exposures can be effective to help students understand computational thinking without computers and tablets.
- Support for Diverse Learning Styles: A single method of communication through technology or gaming control is not for everybody. For instance, diagrams are suitable for visual learners, movement engages kinesthetic learners, and different learning styles accommodate social learners who thrive in collaborative exercises [11].

Unplugged learning does not just allow teachers to include a 'hands on' approach to teaching students but instead creates all sorts of possibilities in terms of obtaining active participation within the target group [10].

7.3.2 Implementing Unplugged Activities in the Classroom

When designing unplugged activities, they must always correspond to the curriculum, tolerate different learning patterns, and encourage interactivity.

Unplugged activities can be implemented in a variety of alternative ways as described below:

- Aligning Activities with Learning Objectives: Educators should select unplugged activities that directly support specific computational thinking skills. For instance, pattern recognition exercises can help students recognize recurring structures, while sorting activities reinforce the principles of organizing data [12].
- Customize Activities for Different Learning Styles: By tailoring activities to accommodate various learning preferences, educators can make the learning experience more impactful. They may use active movement techniques enabling them 'to show' the procedure, whereas it would be more customary for the visual methodological approach to incorporate sketches and diagrams [12].
- Encourage Teamwork and Communication: Many unplugged activities lend themselves well to group work, providing opportunities for students to practice collaboration, cooperation, and learning from each other. This collaboration when solving puzzles or performing tasks ensures that a community is built while students learn how to tackle problems more effectively.
- Inclusion of Individual Thought Patterns, Analysis, and Debriefing: After engaging in the unplugged activities, students should be encouraged to reflect on the process by writing in diaries or through guided questions from educators. This helps students formulate links between the physical and computational components, which is fundamental in this scenario [13].

By very carefully incorporating the unplugged activities, teachers offer students a low-tech and easily understandable method for learning computational thinking. This approach not only enhances student understanding but also equips them with advanced information skills, thus creating a solid base of computational thinking [10].

7.4 Teacher Preparation and Professional Development

Many existing teacher training programs are unable to address the need for educators to be skilled in the implementation of computational thinking, particularly unplugged activities. These gaps may include:

- Lack of Specialized Training: The major gaps in training as highlighted in literature is Type A training wherein there are no specific pedagogical gaps such as computational thinking or unplugged pedagogical activities is which the gap lies [14].
- Resource Constraints: Teacher resource centers and colleges are not equipped and trained to assist teachers to learn practical free technology teaching methods.
- Limited Emphasis on Personalization: A teacher's preparation program pays very little attention to personalizing the issue at hand. Specifically, they pay little to no regard of the various features related to the teacher's learning or personality that are vital for proficient teaching practice adoption [15].
- Time and Curriculum Constraints: Additionally, teachers are reluctant and unable to adopt new teaching methods due to time constraints within the overly ambitious curriculum, leaving very little time for additional training.

These issues underline the need to devise innovative strategies that can give teachers the required skill set and resources to assist in the application of thinking in terms of computation.

7.4.1 Importance of Teacher Readiness for Unplugged Activities

Another important factor is that teachers are the ones who need to be guarded before the execution of thinking in terms of computation using unplugged cross-curricular skills. Well-prepared teachers are more confident, motivated, and knowledgeable, which improves learners' interest and outcomes [6]. Some main causes of this are:

- Enhanced Student Engagement: A teacher able to utilize unplugged strategies is able to provide more engaging and stimulating lessons that students are able to work at and think in terms of computation.
- Adaptability to Diverse Classrooms: A teacher with adequate preparation can implement unplugged activities with the expectation of different learner participation that will assist in capturing more learners' attention [5].
- Long-Term Sustainability: Educators who are adequately trained are more proficient at implementing unplugged activities in curricular areas even after systemic changes in education are adopted.

If teachers are trained appropriately on unplugged pedagogy, they will be able to teach students necessary computational thinking concepts without the use of technology confidently and competently [7].

7.4.2 Strategies for Effective Teacher Preparation

Focused measures need to be put in place to overcome existing obstacles and improve teacher preparedness in computational thinking with the use of unplugged pedagogy. Such measures would be:

- Integrating Learning Styles and Personality-Based Training: Programs can assess teachers' learning styles (visual, auditory, kinesthetic) and MBTI personality types, and trainers can customize the training to enhance teachers' engagement in the training. Such personalization makes it easier for teachers to be trained in what works for them.
- Conducting Workshops and Simulations for Instructional Activities: Workshops that demonstrate unplugged pedagogical approaches to enable teachers to learn and apply them. Participating in such simulated exercises empowers teachers and helps them appreciate how the instruction can be integrated into their classrooms [12].
- Developing Resource Kits for Classroom Implementation: Providing resource kits makes teacher preparation for practical lessons less cumbersome by supplying materials and guides for unplugged activities. Depending on the school's context, these kits may include lesson plans, activity sheets, and tools for easy adaptation and immediate use.
- Ongoing Support and Professional Learning Communities (PLCs): Establishing networks or PLCs where teachers can share best practices, ask questions, and collaborate on new ideas for unplugged computational thinking lessons promotes a sense of community and ongoing motivation [16].

In applying these approaches, teacher training interventions can create a support ecosystem that prepares educators to effectively integrate unplugged computational thinking. With these efforts, teachers can develop students' critical thinking abilities and help create digitally healthy learners with analytical perspectives [13].

7.5 Learning Styles in Education

All individuals have different ways of acquiring and storing information, known as learning styles. While discussions on learning styles have been extensive, modern research suggests that learning preferences are diverse and dynamic [15].

7.5.1 Overview of these Learning Styles (Auditory, Visual, Kinesthetic, Reading/Writing)

Learning styles commonly include four primary categories:

- Visual Learners: Prefer to see and visualize information. Diagrams, illustrations, and visual aids enhance their comprehension and retention.
- Auditory Learners: Learn best through listening. Discussions, lectures, and auditory instructions are effective for these learners [16].
- Kinesthetic Learners: Engage and understand through physical activity and hands-on learning. They benefit from role-playing, movement-based activities, and manipulatives.
- Reading/Writing Learners: Prefer information presented in text. These learners excel when they can read and write notes, lists, and explanations.

Although these categories are somewhat arbitrary, they are useful for addressing instructional strategies in a teacher training scenario [15].

7.5.2 Importance of Recognizing Learning Styles in Teacher Training

Understanding learning styles during teacher development can improve the quality and scope of professional development. Some advantages include:

- Enhanced Retention and Confidence: When training resonates with teachers' preferred learning styles, they are more likely to retain information and feel confident in applying new teaching techniques [17].
- Increased Engagement: A training program that acknowledges different learning styles can reduce monotony, increase engagement, and ensure all participants feel included and supported.
- Better Preparation for Diverse Classrooms: Teachers aware of learning style differences are more likely to understand and accommodate student individual differences, enabling them to employ varied teaching techniques [12].

7.5.3 Assessing Teachers' Learning Styles

This understanding can be developed through tests and self-reflection activities embedded in the training. Instruments such as questionnaires and observational tasks help identify teachers' preferred ways of learning. Once identified, teacher training programs can position activities and materials to advance these findings [9]. For example:

- Video tutorials and illustrated guides would suit the needs of visual learners.
- Learners that are auditory in nature will benefit from conversations, lectures, and even audiotapes of explanations.
- Practitioners, or kinesthetic learners, will benefit from sessions with activities that require hands-on interaction and no use of technology.
- Writing and reading learners will benefit from comprehensive textbooks and procedural manuals.

As teacher training programs try to address and appreciate varying learning preferences, the aim is to make any educational institution work more effectively by appreciating self-differentiation and timeless training [15].

7.6 MBTI Personality Types and Their Impact on Teaching

The most important thing in professional development or self-development is the issue of gaps in understanding people's differences in personalities, particularly in teaching. These gaps stem from learning and teaching people, and meeting students, all of which utilize the Myers–Briggs Type Indicator (MBTI). The same teacher training frameworks concerning MBTI much more effectively implement new computational thinking integration or unplugged activities [16].

7.6.1 Introduction to the Myers–Briggs Type Indicator (MBTI)

The MBTI is one of the most popular tests today and is based on Carl Jung's theory of psychological types. It identifies 16 personality types formed from four opposing dimensions:

- Introversion (I) vs. Extraversion (E): This category determines whether a person exerts energy through internal reflection or through engaging interactions.
- Sensing (S) vs. Intuition (N): Relates to whether information is perceived through direct sensory experience or the abstract patterns.

- Thinking (T) vs. Feeling (F): Determines if decisions are guided primarily by logic or by personal values and relationships.
- Judging (J) vs. Perceiving (P): Suggests a preference for structured decision-making or for adaptability and spontaneity.

These in total allow for the particularity of 16 types, each of which describes the way people engage with their environment and other people [18].

7.6.2 Overview of Different MBTI Personality Types

The 16 personality patterns captured in the MBTI are complex. A few include those listed below:

- ISTJ (The Inspector): Focuses on detail and structure, thus organized and practical, sufficient to bring into effect what has been placed in plan.
- ENFP (The Campaigner): Creative and easily gets excited, works well with others, and is committed to advancing in opportunistic surroundings.
- INTJ (The Architect): Holds independent perspectives and is keen on strategy so as to achieve desired objectives while executing their plans.
- ESFJ (The Caregiver): Genuinely caring and agreeable while prioritizing trustworthy cooperation.

7.6.3 Influence of Personality Types on Teaching Styles and Interactions

MBTI personality types affect classroom management, lesson planning, and student interactions. Key impacts include:
Teaching Styles:

- Extraverted types (e.g., ENFJ, ESFP) often favor interactive, student-centered methods and thrive in energetic environments.
- Introverted types (e.g., ISTJ, INFP) may prefer structured, independent activities and excel in one-on-one interactions [18].

Classroom Management:

- Judging types (e.g., ESTJ, ISFJ) tend to maintain structured routines, fostering consistency and discipline.
- Perceiving types (e.g., ENTP, INFP) might adopt flexible, spontaneous approaches, encouraging creativity and adaptability.

Adaptability to Unplugged Activities:

- Sensing types (e.g., ISTJ, ESTP) may prefer tangible, hands-on activities that align with unplugged computational thinking methods.
- Intuitive types (e.g., ENFP, INFJ) may gravitate toward conceptual or open-ended activities, helping students explore ideas creatively [18].

Understanding these differences helps training programs support teachers in leveraging their strengths while developing well-rounded instructional skills. Recognizing and adapting to diverse MBTI types fosters a supportive environment that enhances educators' ability to connect with students and facilitate engaging learning experiences [19].

7.7 Integrating Learning Styles and MBTI into Teacher Training

To promote clarity and accommodate diverse teaching styles, the following framework is organized into distinct, numbered steps and concise guidelines.

Personalized teacher training that considers individual learning styles and personality types can enhance educators' preparation, particularly when teaching computational thinking through unplugged activities. By tailoring training programs to align with each teacher's preferences and strengths, training becomes more relevant, engaging, and effective.

To ensure different teaching approaches are clear and effective, the provided framework has been broken down into easy-to-follow steps and instructions.

Unplugged training activities can be taught more elegantly if personalized teacher training aligns with the different personalities and learning styles of the educators. Training becomes more effective and interesting when it is tailored to the needs of the trainer through their preferences and strengths.

7.7.1 Tailoring Teacher Training Programs to Individual Needs

One way of solving the problem of the disparity between a teacher's learning style and personality is being able to develop modules that target all their needs. For instance:

- Customized Training Modules: Modules can be designed in multiple formats—such as visual aids, interactive discussions, and hands-on practice—allowing each teacher to engage in a manner that suits their preferred learning style [15].

- Adaptive Support: Trainers can use MBTI insights to provide targeted guidance, such as offering more collaborative opportunities to extroverted teachers while providing self-paced activities for introverted teachers.
- Feedback Mechanisms: Personalized feedback tailored to each teacher's learning preferences can encourage improvement in areas aligned with their strengths and growth in areas needing development.

This framework recognizes that teachers are not uniform; it is flexible enough to be adapted to each teacher's unique instructional style and personality type.

7.7.2 Developing Personalized Teaching Strategies Based on Learning Styles and Personality Types

By integrating learning styles and personality types into training, teachers can develop personalized strategies that they can apply directly in their classrooms. Examples include:

- For Visual Learners and Sensing Types: Teachers may develop visually rich lesson plans, utilizing diagrams, charts, and tangible materials that enhance computational thinking through clear, sensory-based learning paths.
- For Auditory Learners and Feeling Types: Teachers can craft interactive, discussion-based activities that foster an emotionally supportive environment where students feel encouraged to explore ideas collaboratively [20].
- For Kinesthetic Learners and Perceiving Types: Unplugged activities can include movement, role-playing, and hands-on problem-solving to help kinesthetic learners grasp computational concepts through active engagement.

7.7.3 Integrating Learning Styles and MBTI into Teacher Training

Personalized teacher training that considers individual learning styles and personality types can enhance educators' preparation, particularly when teaching computational thinking through unplugged activities. By tailoring training programs to align with each teacher's preferences and strengths, training becomes more relevant, engaging, and effective.

Case Studies or Examples of Successful Implementation

CASE STUDY 1

Enhancing Teacher Training Through Psychometric Assessment and Data-Driven Evaluation [21]

INTRODUCTION

Problem Statement: Despite numerous teacher training initiatives, a persistent challenge remains in accurately assessing teachers' self-awareness, reading comprehension, and overall academic competency to deliver targeted professional development. This case study addresses this challenge by applying the Sudakshta psychometric framework and dataset to provide data-driven insights and personalized feedback, thereby enhancing teacher performance and effectiveness.

Many teacher training initiatives fail to accurately assess teachers' self-awareness and academic competencies, leading to generic feedback that does not address individual needs. The system combines self-assessment, multiple-choice question (MCQ) testing, and reading comprehension analysis to deliver personalized training and actionable feedback for teachers (Figure 7.2). The goal is to improve

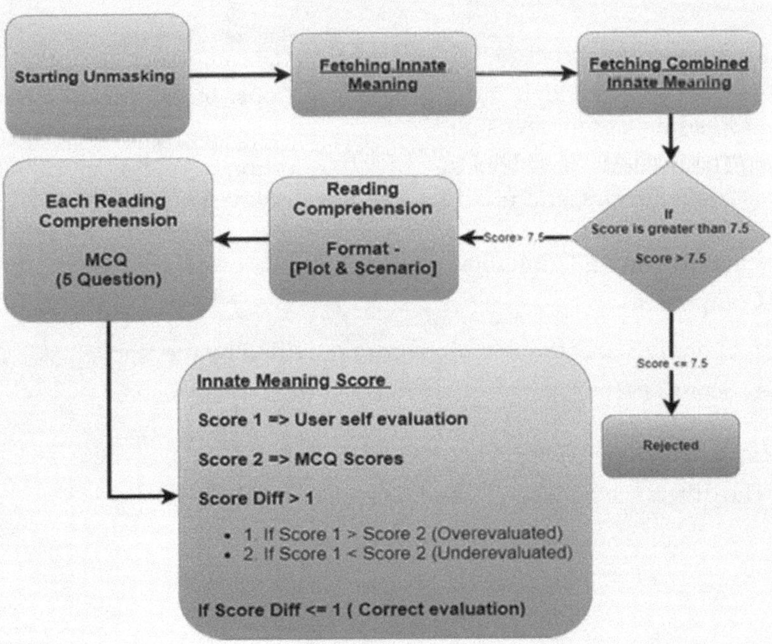

FIGURE 7.2
Innate meaning assessment process.

self-awareness, reading comprehension, and overall academic competency. This evaluation follows the Sudakshta psychometric framework and utilizes the Sudakshta dataset to assess and improve teacher performance.

OBJECTIVES

1. Assess Self-Awareness: To determine the accuracy of teachers' self-assessments compared to their objective performance in the "Innate Meaning" section.
2. Measure Reading Comprehension: To evaluate reading comprehension through structured scenarios and questions.
3. Evaluate Overall Competency: To calculate a comprehensive score that highlights teachers' strengths and areas that need improvement.
4. Deliver Personalized Feedback: To provide customized recommendations for teachers, enhancing their professional development and teaching effectiveness.

METHODOLOGY

1. Complete Test Structure:
   - The test is divided into two main sections: Innate Meaning and Reading Comprehension.
   - The overall test consists of 60 questions, assessing 12 sub competencies, and provides a comprehensive evaluation score.

2. Innate Meaning Evaluation:

   Components:

   - Score 1 (S1): Self-evaluation score provided by the teacher.
   - Score 2 (S2): Objective score derived from MCQ testing.

   Score Difference (Diff) Analysis:

   The difference between S1 and S2 is calculated: $\text{Diff} = |S1 - S2|$

Evaluation Types:

- Overevaluated: If Diff > 1 and S1 > S2, indicates the teacher has overestimated their performance.
- Underevaluated: If Diff > 1 and S1 < S2, indicates the teacher has underestimated their performance.
- Correct Evaluation: If Diff ≤ 1, indicates accurate self-awareness.

3. Combined Innate Meaning Score (C):
   - The Innate Meaning section has a maximum score of 12.
   - If the Combined Score (C) > 7.5, it is flagged for further review, suggesting a need for additional training or coaching.
4. Reading Comprehension Evaluation:
   - Teachers answer five sets of MCQs, each following a scenario-based format.
   - Each question set contributes to the Reading Comprehension Score (R), with a maximum score of 60.
5. Overall Test Score (T):
   - Calculated as: T = S2 + R = S2 + R
   - The overall score evaluates the teacher's academic and comprehension competency, with a maximum score of 120.

ANALYSIS AND INTERPRETATION BASED ON TABLE 7.1

TABLE 7.1
Sudakshta Dataset and the Evaluation Metrics

Teacher ID	Self-Evaluation Score (S1)	MCQ Score (S2)	Diff	Evaluation Type	Combined Score (C)	Flag (C > 7.5)	Ri Scores (5 sets)	Reading Comp. Score (R)	Total Test Score (T = S2 + R)
T001	8	7	1	Correct Evaluation	7.5	No	[10, 8, 9, 11, 10]	48	55
T002	6	4	2	Underevaluated	8	Yes	[7, 9, 10, 8, 9]	43	47
T003	5	6	1	Correct Evaluation	6.5	No	[8, 7, 6, 9, 8]	38	44
T004	9	7	2	Overevaluated	8.5	Yes	[11, 10, 10, 11, 12]	54	61

1. Self-Assessment Accuracy:
    - Overevaluated Cases: Teachers such as T004 overestimated their performance. Personalized feedback includes strategies to improve self-awareness and align self-perception with actual performance.
    - Underevaluated Cases: Teachers like T002 underestimated their abilities. Feedback includes confidence-building exercises to help them recognize their strengths.
    - Correct Evaluation: Teachers T001 and T003 showed accurate self-assessment, demonstrating good self-awareness and self-evaluation skills.
2. Combined Innate Meaning Score:
    - Teachers with a score greater than 7.5 (e.g., T002 and T004) are flagged for further review. These teachers may require targeted training to improve in specific areas.
3. Reading Comprehension Score:
    - Reading comprehension scores vary across teachers. For instance, T002 has a lower comprehension score, indicating the need for additional support in reading and comprehension skills.
    - Teachers with higher scores (e.g., T004) show strong comprehension abilities but may need to refine other competencies.
4. Total Test Score (T):
    - The total score provides a comprehensive assessment of the teacher's academic and comprehension skills. Lower overall scores (e.g., T002) highlight the need for further training and support.

PERSONALIZED TRAINING AND FEEDBACK

Using advanced AI tools like GPT-3.5, personalized feedback is generated for each teacher:

- Overevaluated Teachers (e.g., T004): Recommended self-awareness workshops, mindfulness training, and peer feedback sessions.
- Underevaluated Teachers (e.g., T002): Confidence-boosting exercises, strengths-focused coaching, and affirmations to build self-esteem.

- Reading Comprehension Support: Interactive reading programs, structured reading techniques, and comprehension-enhancing exercises.
- General Training Suggestions: Use of visual aids, adaptive learning platforms, and scenario-based teaching modules tailored to individual needs.

Overall, this case study demonstrates the efficacy of a data-driven evaluation framework for teacher training, using the Sudakshta psychometric framework and the Sudakshta dataset. By integrating self-assessment, objective testing, and reading comprehension analysis, the system provides detailed insights and personalized feedback to enhance teachers' professional development. The structured approach ensures that teachers receive targeted support, ultimately improving their teaching competencies and self-awareness.

CASE STUDY 2

Empowering Workforce Development with Sudakshta's Emerging Psychometric Framework [22]

INTRODUCTION

Problem Statement: One of the significant challenges that organizations face is accurately assessing and aligning employee competencies with relevant actions and behaviors. Traditional assessments often fall short of offering the granular insights needed for meaningful development. To address this challenge, Sudakshta aimed to test its innovative Jumbled Competency (JC) Model, which promises a structured and insightful approach to competency assessment (see Figure 7.3).

Sudakshta Inc., an emerging startup specializing in psychometric assessments and competency-based training, was founded with a vision to revolutionize how organizations evaluate and enhance employee performance. As a new player in the field, Sudakshta has developed a unique Psychometric Framework designed to provide deeper insights into employee competencies, paving the way for more data-driven and effective workforce development strategies (see Figure 7.3).

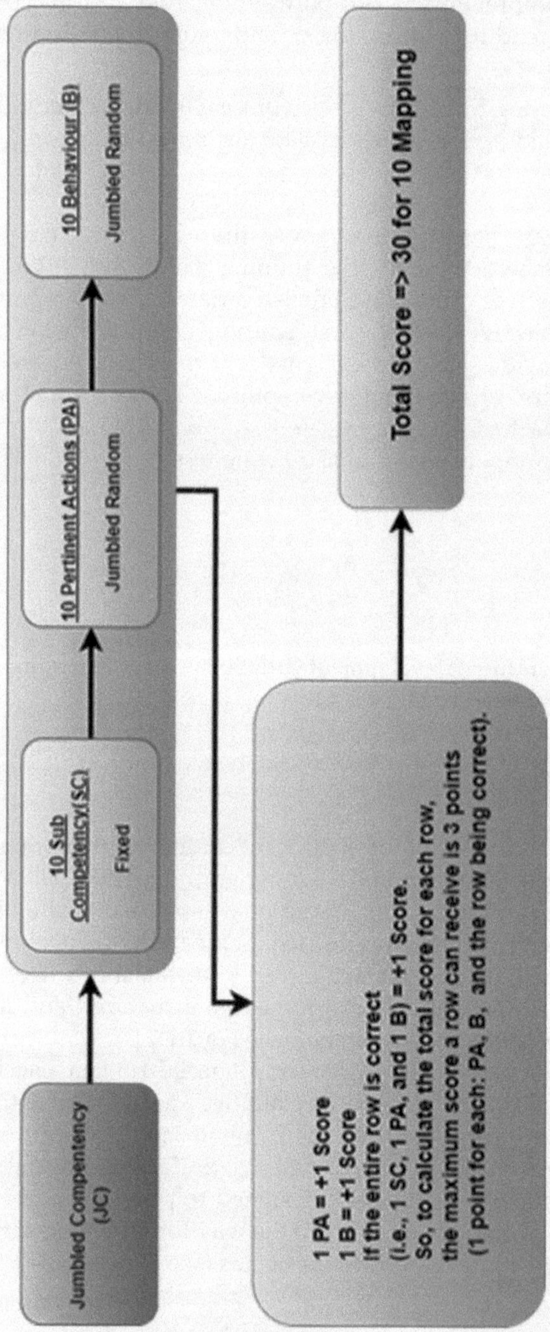

FIGURE 7.3
Jumbled Competency (JC) model: aligning sub-competencies, actions, and behaviors.

OBJECTIVES

1. Psychometric Precision: To establish a reliable and valid assessment framework that meets high psychometric standards, despite being developed by an emerging startup.
2. Identifying Development Needs: To uncover competency gaps and actionable insights for targeted employee development.
3. Scalability for Growth: To create a model that could be easily scaled and adapted as Sudakshta grows and expands its reach.

IMPLEMENTATION OF THE SUDAKSHTA JUMBLED COMPETENCY MODEL

Step 1: Structuring the Competency Framework

Sudakshta's team identified 10 core subcompetencies (SC) that are critical for high performance in a modern workplace. These subcompetencies were carefully chosen to align with key attributes like problem-solving, effective communication, and emotional intelligence.

- Example of Subcompetency: Adaptability
  - Pertinent Action: Effectively managing unexpected changes in a project.
  - Behavior: Demonstrating a proactive attitude when dealing with shifting priorities.

Step 2: Designing the Jumbled Competency Assessment

To challenge participants' understanding and accurately measure their alignment of competencies, the JC Model was designed as follows:

- 10 Subcompetencies (SC): Fixed order to maintain consistency.
- 10 Pertinent Actions (PA): Randomly jumbled, requiring participants to identify which action best matches each subcompetency.
- 10 Behaviors (B): Also presented in a random order, adding a layer of complexity to the assessment.

Step 3: Innovative Scoring System

Sudakshta's scoring method ensures precision and fairness:

- 1 Pertinent Action Correct (PA) = +1 Point
- 1 Behavior Correct (B) = +1 Point
- Entire Row Correct (SC + PA + B) = Additional +1 Point
- Maximum Score: 30 points (10 rows × 3 points per row)

This scoring system is built to encourage accuracy and reward comprehensive understanding, adhering to high psychometric standards.

PILOT EXECUTION AND DATA COLLECTION

As a startup, Sudakshta conducted a pilot assessment with a client in the tech industry, engaging 100 employees in the process. The pilot was designed to test the robustness and scalability of the model while gathering valuable feedback from the client.

Results and Analysis

The assessment yielded insightful data, critical for a growing company like Sudakshta:

1. Score Distribution:
    - Average score: 21 out of 30 (70%).
    - High performers (top 10%): Scored between 26 and 30 points, showing strong alignment between competencies, actions, and behaviors.
    - Low performers (bottom 10%): Scored between 15 NS 18 points, revealing key areas for improvement.
2. Key Competency Gaps:
    - Many participants struggled with linking behaviors related to adaptability and problem-solving.
    - There was a notable difficulty in associating actions involving leadership with the correct subcompetence.
3. Psychometric Evaluation:
    - Reliability: The model demonstrated a promising reliability score with a Cronbach's alpha of 0.85, showing that the JC model is robust even at the early stage of development.
    - Validity: Preliminary expert reviews affirmed the content validity of the framework, suggesting that it has strong potential for future enhancement.

STRATEGIC RECOMMENDATIONS

1. Focused Training Initiatives: Sudakshta suggested the development of targeted workshop schemes based on adaptability, leadership, and strategic thinking which are more beneficial for greater results.

2. Behavioral Coaching: Sudakshta positioned a new offering for an entrepreneurial company that helps participants in areas where they are less able to cope, through personalized coaching service.
3. Iterative Development: The data collected in the pilot study were very important. Sudakshta intends to refine the framework and design an even more sophisticated jumbled randomization algorithm, which will allow a greater degree of accuracy.

IMPACT AND EARLY SUCCESS

1. Sudakshta's JC Model has made great strides toward early clientele retention. Sudakshta has incorporated innovative features and adopted a holistic approach, which has drastically improved their impression in the market.
2. A recent report stated that the assessment brought to light many aspects that were never considered before and helped in building more robust employee development programs.
3. Modular components of JC Model help in expansion within newer business domains seamlessly, making it easier for Sudakshta to target untapped areas with minimal effort.

LESSONS LEARNED

1. The Power of Innovation: Sudakshta is distinguished from other assessment providers by its offering new jumbled competencies with deep insight into the complexities of the field.
2. Early Validation: The pilot study revealed the feasibility of creating a robust psychometric tool by a startup, which could be achieved with the right approach and methodology.
3. Early Validation: The pilot study confirmed that even a startup could create a psychometrically robust tool, given the right approach and methodology.

Above all, as a new emerging competitor in the field of psychometrics, Sudakshta Inc was able to record an insightful experience that was primarily driven by innovation and adaptability. In delivering insightful evaluation of the competencies, Sudakshta was able to successfully use The Sudakshta Psychometric Framework and its Jumbled Competency Model. With growth in its stature, the company stands ready to use the industry-advanced techniques in psychometry to make meaningful transformations in the sphere of workforce development.

SUMMARY OF KEY INFERENCES

To aid in the comprehension and dissemination of the results quite simplistically, Table 7.2 contains the most pertinent findings which stemmed from the study.

TABLE 7.2

Key Inferences of Case Study 2

Problem Being Addressed	Key Findings	Implications for Teacher Preparation
Difficulty in accurately assessing and aligning employee competencies with corresponding actions and behaviors	The JC Model demonstrated robust reliability (Cronbach's alpha = 0.85) and revealed specific competency gaps in areas such as adaptability and leadership	Similar psychometric frameworks can be applied in teacher preparation to identify competency gaps, enabling targeted professional development and tailored training interventions

7.8 Leveraging Technology in Teacher Preparation

AI is becoming a new frontier within education with its enormous capacity to redefine how trainers deliver teacher training programs. Machine learning and artificial intelligence have tremendous potential in reshaping education for the better, and teacher preparation programs are beginning to shift resources toward these advanced technologies. AI models can be used by teacher training programs to evaluate and assist a teacher's unique teaching style, and in doing so, they will be able to target a teacher's specific learning style and personality trait [17]. This section investigates the applications of machine learning and AI within education by focusing on LSTM, LinearSVC, and GPT-3.5 models that deliver the growing applicability of these technologies to provide unique education to instructor trainees.

7.8.1 Role of Machine Learning and AI in Education

The possibilities of AI and machine learning in education are far-reaching in personalizing adaptive feedback, and identifying preferences and content delivery. In teacher preparation, ML and AI can be applied for:

- Recognition of Learning Styles and Personality Type Patterns: Memories of teachers' ability classification based on individual differences in learning styles and MBTI personality types can be captured using artificial intelligence [4].
- Delivering Customized Training: AI technologies can identify distinct learning needs, and as a result, they can tailor training modules to specific strengths of teachers to improve their confidence and motivation.
- Deliver Real-Time Feedback: AI-driven platforms can provide instant, specific feedback to teachers during training, helping them refine their skills in real time [8].

In this way, AI helps create a more engaging experience in which each teacher feels supported and catered for during the learning experience.

7.8.2 Overview of the Long Short-Term Memory (LSTM) Model for Learning Style Classification

LSTM networks are a type of recurrent neural network. LSTMs are ideal for tracking these learning styles because they can remember information over long sequences which makes them excellently capable of analyzing text data in detail. In this training model, teachers input their preferred educational methods, and the LSTM classifies them into visual, auditory, kinesthetic, or reading/writing learners. This model is trained on a ready-made labeled dataset that has been preprocessed by removing punctuation and lowercasing letters as well as tokenizing. Such classification permits adaptive training programs to be developed that can help tailor materials and the teaching approach to each of the learning styles, thus improving the overall preparation to the teacher's expectations [17, 18].

7.8.3 Utilizing LinearSVC for MBTI Personality Classification

The Linear Support Vector Classifier (LinearSVC) is also an AI model for classifying the Myers–Briggs Type Indicator personality types. LinearSVC is a quick and effective text classification model, which makes it appropriate for classifying personality questionnaire responses. Based on patterns within the provided answers, this model utilizes labeled datasets to classify each teacher surveyed into one of the 16 MBTI personality types (e.g. ENFP, ISTJ).

The classification from MBTI can, in turn, be utilized to revise and adapt the instructional materials and strategies to help every employed educator leverage their natural characteristics and qualities. Take, for example, an ENFP teacher: He may want to use collaborative, intensive activities instead of lectures. In contrast, an ISTJ instructor may opt for self-paced, structured modules [15].

7.8.4 Implementation of GPT-3.5 for Personalized Teacher Guidance

GTP-3.5 provides special personalization for each and every teacher, along with motivation and practical examples that are hard to provide in traditional classrooms. By retraining the AI model on business datasets relating to classroom management as well as strategy, it is able to have real-time conversations with teachers who can be guided in a way that they prefer, depending on their MBTI type and learning style [8].

Therefore, a supporting and empathetic speech would make an intuitive, feeling-oriented teacher more confident in their methods. In another scenario, a visual learner could be given diagrams, examples, and explanations while they learn new methods of teaching [12].

GTP-3.5 combined with AI leads teachers into being more adaptive and effective, while they are being guided through the learning process. Since machine learning can be implemented into aligning styles of learning with LSTMs, while incorporating LinearSVC for MBTI personality classification and GPT-3.5, it becomes easy to provide this level of personalization for teacher training.

Through the use of LSTM for learning style identification, LinearSVC with integrated MBTI for personality type classification, along with GPT-3.5 to assist in a dynamic and personalized way, teacher training programs can be highly customized. With technology-integrated help in making teachers to be more prepared and trained, it becomes responsive, pertinent, and more fitting to the multifaceted requirements of teachers today [23].

7.9 Practical Implementation Framework

In attempting to make an effective teacher training program that focuses on these learning styles and MBTI personality types, the framework needs to be detailed and comprehensive. This framework details the approach educational institutions and training providers need to take in creating specialized, adaptive training sessions that all teachers with varying learning styles and personalities will be able to benefit from. This part also provides example unplugged activities that are designed based on the different learning styles and personality types and guidance on how to assess the effectiveness of such program.

7.9.1 Step-by-Step Plan for Incorporating Learning Styles and MBTI into Teacher Training

1. Initial Assessment of Learning Styles and Personality Types
   - o Conduct an initial assessment to determine each teacher's learning style (visual, auditory, kinesthetics, or reading/writing) and MBTI

personality type. This can be done through a survey or questionnaire analyzed by machine learning models like LSTM for learning styles and LinearSVC for MBTI classification.

2. Customizing Training Materials

   Use assessment data to create or adjust training materials that align with each teacher's identified learning style and personality type. For example:

   - o Visual learners could receive materials with infographics and diagrams.
   - o Auditory learners might benefit from audio explanations or lectures.
   - o Introverted personality types could be assigned reflective, independent tasks, while extroverted teachers might engage in collaborative activities [16].

3. Personalized Instruction Using GPT-3.5

   - o Implement GPT-3.5 to provide personalized feedback, explanations, and examples throughout the training process. By adjusting responses according to each teacher's learning preferences and personality, GPT-3.5 can enhance engagement and comprehension [24].

4. Design and Delivery of Unplugged Activities

   - o Plan unplugged activities that cater to different learning styles and personality types. Include a variety of exercises, such as collaborative problem-solving for extroverted learners or structured step-by-step tasks for kinesthetics learners, to ensure that all personality types and preferences are engaged.

5. Continuous Feedback and Adjustment

   - o Use formative assessment, verification, and teacher formative feedback reports to enhance every teacher's progress tracking. Change activities or resources or teaching techniques, if necessary, toward achieving the pre-set individual feedback and learning outcomes [23].

6. Final Evaluation and Reflection

   - o After the training exercises, a final evaluation should be conducted in order to assess the learning outcomes as well as gains made in computational thinking instruction. Teachers should be able to share their views with regard to the training methods and the practice of personalized training.

7.9.2 Examples of Unplugged Activities Tailored to Different Learning Styles and Personality Types

- Visual Learners (ISTJ, ISFJ): Use flowcharts or visual storytelling activities that outline computational steps in problem-solving.
- Auditory Learners (ENFP, ESFP): Include discussion-based activities and storytelling to explain computational concepts aloud, followed by group reflection.
- Kinesthetic Learners (ESTP, ISTP): Organize hands-on activities like using physical blocks to represent algorithms, allowing learners to engage with the material tangibly.
- Reading/Writing Learners (INTJ, INFJ): Provide problem-solving scenarios in written form, encouraging learners to analyze, annotate, and devise solutions [15].

7.9.3 Evaluation Metrics for Success

To implement such an individualized strategy with maximum efficacy, it is advisable to determine the relevant indicators for evaluation purposes. The following metrics will assist in evaluating the effectiveness of the training program:

- Learning Outcomes Improvement: Evaluate teachers' understanding and application of computational thinking concepts before and after training.
- Engagement and Participation Levels: Measure the levels of active engagement in unplugged activities, noting any correlation between tailored activities and increased participation.
- Personalized Support Effectiveness: Use feedback forms to assess the perceived usefulness of personalized support provided by AI tools like GPT-3.5.
- Self-Reported Confidence: Gauge teachers' confidence in teaching computational thinking concepts, particularly in implementing unplugged activities in the classroom.

This practical framework guides the development of a highly individualized teacher training program that prepares educators to effectively integrate computational thinking through unplugged activities. By tailoring the experience to each teacher's learning style and personality, institutions can foster a more inclusive, engaging, and impactful learning environment for teachers and students alike [24].

7.10 Conclusion

In this chapter, we detail an approach to comprehensive teacher preparation dedicated to teaching computational thinking that also incorporated learning styles and MBTI personality types. Knowing what each teacher needs within training programs enables targeting other forms of support that improve the effectiveness of unplugged activities as one of the ways of fostering computational thinking across the curriculum. Such a strategy not only prepares teachers in terms of better grasp and articulation of computational ideas but also creates a culture of flexibility and inclusivity in the classroom.

From an analysis of the case studies, it can be noted that the proposed model had a positive effect on self-assessment accuracy and competency development of the teachers. The results evaluation indicates that the use of AI tools such as the learning style classification LSTM, personality type classification LinearSVC of MBTI, and personalized AI with GPT 3.5 leads to improvement in teacher output and focused professional growth. Highlights of this evaluation include the model's remarkable accuracy, the capacity to identify key issues to be addressed about the teachers' performance, and the overall proposition to revolutionize training models for teachers. From a methodical point of view, this study provides added value by proposing a new approach that combines cutting-edge AI with personalized instruction that is more efficient in preparing teachers.

Main Ideas: Computational Thinking and Unplugged Activities: We defined the term computational thinking and explained the concept of unplugged activities as one way to teach basic skills without using technology. Need for Teacher Preparation: We were able to solve the problems as a teacher trainer on how to prepare teachers who need to teach ICT and focus on computational thinking pedagogy. Learning Styles and MBTI Personality Types: Once the learning styles and common personality traits of the teachers are revealed, training programs can be designed to their specific needs, thus improving the teaching and learning processes. Application of Technology in Teacher Education: ML and AI, or more specifically, LSTM for learning style identification and GPT-3.5 for tailored advice, provide a new efficient and versatile approach for teacher education. Practical Framework for Implementation: A coherent, step-by-step framework gives educational institutions the ability to work on these concepts with a clear agenda and set guidelines on how to measure training success.

7.11 Future Directions for Research and Practice

An example of future work could include how personalized conventional training approaches applied over an extended period would influence teacher effectiveness. In addition, the need to design more unplugged activities for different learning and personality types is more likely to create even greater diversity in instructional strategies. Further investigation into personalized AI tools, generative models, and adaptive learning systems would facilitate continuous improvement starting from primary schools and extending to higher education institutions.

7.12 Call to Action for Educators and Institutions

To aid with the preparation of future educators, novel technologies like cloud-stored apps and software can be utilized at the same time. Taking into account different learning styles, personality types, and AI tools would enable teachers to effectively teach computational thinking to students as a skill that is fundamental in the 21st century.

References

1. Li, F., Wang, X., He, X., Cheng, L., & Wang, Y. (2022). The effectiveness of unplugged activities and programming exercises in computational thinking education: A Meta-analysis. *Education and Information Technologies*, *27*(6), 7993-8013.
2. Domínguez, L. G. I., Clavijo, F. V., Robles-Gómez, A., & Pastor-Vargas, R. (2024). A Sustainable Educational Tool for Engineering Education Based on Learning Styles, AI, and Neural Networks Aligning with the UN 2030 Agenda for Sustainable Development. *Sustainability*, *16*(20), 1-24.
3. Štajner, S., & Yenikent, S. (2021, September). How to Obtain Reliable Labels for MBTI Classification from Texts?. In *Proceedings of the International Conference on Recent Advances in Natural Language Processing (RANLP 2021)* (pp. 1360-1368).
4. Simkus, J. (2024). "Myers-Briggs Type Indicator (MBTI): 16 personality types." *Simply Psychology*. Retrieved from https://www.simplypsychology.org/the-myers-briggs-type-indicator.html
5. Wing, J. M. (2006). Computational thinking. *Communications of the ACM*, *49*(3), 33-35.
6. Grover, S., & Pea, R. (2013). Computational thinking in K–12: A review of the state of the field. Educational researcher, *42*(1), 38-43.
7. Brennan, K., & Resnick, M. (2012, April). New frameworks for studying and assessing the development of computational thinking. In *Proceedings of the 2012 annual meeting of the American educational research association, Vancouver, Canada* (Vol. 1, p. 25).

8. Bers, M. U., Flannery, L., Kazakoff, E. R., & Sullivan, A. (2014). Computational thinking and tinkering: Exploration of an early childhood robotics curriculum. Computers & education, *72*, 145-157.
9. Liu, Z., Gearty, Z., Richard, E., Orrill, C. H., Kayumova, S., & Balasubramanian, R. (2024). Bringing computational thinking into classrooms: a systematic review on supporting teachers in integrating computational thinking into K-12 classrooms. International Journal of STEM Education, *11*(1), 51.
10. Payne, L., Tawfik, A., & Olney, A. M. (2022). Computational thinking in education: Past and present. TechTrends, *66*(5), 745-747.
11. Rao, T. S. S., & Bhagat, K. K. (2024). Computational thinking for the digital age: a systematic review of tools, pedagogical strategies, and assessment practices. *Educational technology research and development*, *72*(4), 1893-1924.
12. Wang, C., Shen, J., & Chao, J. (2022). Integrating computational thinking in STEM education: A literature review. *International Journal of Science and Mathematics Education*, *20*(8), 1949-1972.
13. Cabrera, L., Ketelhut, D. J., Mills, K., Killen, H., Coenraad, M., Byrne, V. L., & Plane, J. D. (2024). Designing a framework for teachers' integration of computational thinking into elementary science. *Journal of Research in Science Teaching*, *61*(6), 1326-1361.
14. Su, J., & Yang, W. (2023). A systematic review of integrating computational thinking in early childhood education. *Computers and Education Open*, *4*, 100122.
15. Cassidy*, S. (2004). Learning styles: An overview of theories, models, and measures. *Educational psychology*, *24*(4), 419-444.
16. Myers, I. B. (1962). *The myers-briggs type indicator* (Vol. 34). Palo Alto, CA: Consulting Psychologists Press.
17. Kamalov, F., Santandreu Calonge, D., & Gurrib, I. (2023). New era of artificial intelligence in education: Towards a sustainable multifaceted revolution. *Sustainability*, *15*(16), 12451.
18. Thompson, R. (2016). *Using the MBTI in Education in the Way It Was Designed*. The Myers-Briggs Company. Retrieved from www.ncda.org/aws/NCDA/pt/sd/news_article/131123/_self/layout_details/false
19. Kamal, A., & Radhakrishnan, S. (2019). Individual learning preferences based on personality traits in an E-learning scenario. Education and Information Technologies, *24*(1), 407-435.
20. Rosati, P., Dean, R. K., & Rodman, S. M. (1988). A study of the relationship between students' learning styles and instructors' lecture styles. IEEE transactions on education, *31*(3), 208-212.
21. Sudakshta. "Case Study 1." Retrieved February 7, 2025, from https://sudakshta.com/
22. Sudakshta. "Case Study 2." Retrieved February 7, 2025, from https://sudakshta.com/
23. Prasad, S., Mohapatra, S., Rahman, M. R., & Puniyani, A. (2022). Investor sentiment index: a systematic review. *International Journal of Financial Studies*, *11*(1), 6.
24. Brown, T., Mann, B., Ryder, N., Subbiah, M., Kaplan, J. D., Dhariwal, P., ... & Amodei, D. (2020). Language models are few-shot learners. *Advances in Neural Information Processing Systems*, *33*, 1877-1901.

8

Revolutionizing AI to Break Down Barriers in Education for Children with Special Needs

Srishti Singh Kaira, Shreya Tyagi and Taranjeet Duggal

8.1 Children with Special Needs (CWSN): An Overview

Learners with special needs refers to a category of students with certain physical, psychological, and social difficulties and conditions that affect their ability to learn in the mainstream classroom (Mehta, 2023). The 2011 Census of India highlighted upon the prevalence of disabilities in the country, with over 2.68 million children aged between 0–19 years having some form of disability (Datta, 2023). These include:

a. Specific Learning Disabilities (SLDs) such as dyslexia: a neurodevelopmental condition that leads to difficulties (or inabilities) to read, decode, and spell words (Snowling et al., 2020); dyscalculia: difficulty in doing simple and complex arithmetic operations and mathematical tasks (Haberstroh & Schulte-Körne, 2019); to name a few.

b. Autism Spectrum Disorders (ASD), characterized by difficultly in social interactions, fixed patterns of repetitive behaviours, troubles in adjusting to social situations, etc. that make it difficult for educators to connect with such students and help them learn in a mainstream classroom setting (Autism Spectrum Disorder, n.d.).

c. Attention Deficit Hyperactivity Disorder (ADHD), marked by symptoms of inattention, impulsivity, and hyperactivity that often make it challenging for the learner to sit and attend lectures, hampering their academic and social lives (Attention Deficit/Hyperactivity Disorder, n.d.).

d. Physical disabilities like hearing and visual impairments, motor disabilities and diseases like muscular dystrophy, cerebral palsy, arthritis, or partial paralysis due to any of these conditions, etc. These difficulties require additional support system in the classrooms, the absence

 DOI: 10.1201/9781003560814-8

of which can cause feelings of inferiority, isolation, marginalization, and missed opportunities of learning (Mehta, 2023).

e. Other cognitive deficits such as intellectual disabilities, mental retardation, and brain-related injuries leading to academic difficulties and interpersonal challenges.

Learners with any of the aforementioned challenges often face a difficult time adapting to their social world. Some of these challenges include unwelcoming attitude by peers due to lack of awareness around such disabilities, lack of appropriate resources in schools that support these children in their academic journey, lack of trained special educators, inconsistencies with the implementation of services and educational provisions done by the legislative authorities, the presence of more disciplinary actions than constructive solutions and interventions, and lack of collaborative practices that bring together parents, schools, learners, and special educators (Mehta, 2023; Dass, 2024).

It is worthwhile to note here that, if not all, a few of these concerns can be efficiently dealt with through the implementation of artificial intelligence (AI) in the education settings. The following sections present an over-arching view of how AI has been proven to be beneficial for learners with difficulties, and what possibilities lie ahead for the education realm.

8.2 Artificial Intelligence in Education

AI-powered platforms create customized learning pathways according to the student's potential. Such platforms change the difficulty of the content in real time without hampering the flow of the student. So, for example, if a student is good in one subject the system would ask that student difficult questions each time he/she answers correctly, whereas, on the other hand, students who are particularly disinterested or weak in that subject could seek assistance from the system (Store, 2024; Be10x, 2024). Adaptive learning platforms have shown great results in encouraging student self-regulation by providing data-derived feedback and personalized recommendations. These tools are experts in understanding a student's behaviour and offer insights for enhanced learning (Afzaal et al., 2021).

Targeted Pacing and Content Generation: AI assistance in education enables students to learn at their own pace. The AI learns the patterns of the student like how much time they take to complete an assignment or monitor their performance metrics through quiz results and then modifies its pace of delivering instructions so that the student is neither rushed into learning nor left behind (Stefanic, 2024). The systems that adapt instructions according to the individual are called adaptive instructional systems (Roessingh et al., 2019).

Catering to Various Learning Styles: Education does not have to always be written or read, it should be accessible to all learning styles. AI helps in achieving this goal by providing visual, audio, or even kinaesthetic ways of learning. For example, some students might be visual learners, therefore, AI systems would generate flowcharts and diagrams for detailed explanations, whereas some might be audio learners, and for them these systems can also create audio lessons for enhanced engagement (Atoui, 2024). The following are some tools that help in adapting to new learning styles:

- o *WELSA (Web-based Educational System with Learning Style Adaptation)*: This is an online learning platform that changes the content according to the student's learning style.
- o *TSAL (Two-Source Adaptive Learning System)*: TSAL works similarly to WELSA, the only difference is that it gathers information from two different sources like grades in the assignment, or class performance marks, etc., and the time it takes to finish it (Pardamean et al., 2022).

Grading: AI tools like Gradescope and Cograder can help educators in automatic and efficient grading while reducing their efforts. These tools can divide the responses of the students into categories and help in grading similar answers (AI Tools for Teacher Grading — TOM DACCORD, n.d.). Meanwhile, tools like Vexis and Timely Grader also provide comments/feedback along with grades. These comments are made based on predefined rubrics allowing teachers to finalize assessments quickly (AI Grading Tool for Teachers | CoGrader | Use Your Own Rubric, n.d.).

8.3 Artificial Intelligence in Promoting Inclusivity for Children with Special Needs

AI has the potential to change the educational framework of our society by not just indulging in the education of the typical population but also lending a helping hand to the divergent and special needs population as well. With the help of AI, educators can help special needs children to compete with mainstream students.

- *Recognizing the Early Signs:* AI can recognize the early symptoms of learning disabilities like dyslexia by analysing the behavioural patterns that point towards specific areas that might cause stress or cognitive overload (Jet Learn, 2024).

- *Better Communication*: Tools like AAC (Augmentative and Alternative Communication) help students to express themselves freely. We all know that some children with special needs struggle in communication, therefore, the use of this device as an AI would not only bridge the gap by suggesting words and phrases that the student is likely to use but also help students who are completely mute or unable to comprehend words (Admin, 2024).
- *Visual Aid:* For visually impaired students, AI can assist in converting text to speech, and describe images, diagrams, and graphs through speech. It also has the ability to magnify images without distorting them for better clarity (Rajagopal et al., 2023; Jet Learn, 2024).
- *Captioning and Translations:* Students with hearing impairment can also benefit from AI. This allows transcribing of speech to text in real time and every video or movie would have automated captions. With the increased need of an hour, AI is being developed to recognize sign language for uninterrupted interaction between students and teachers (Rajagopal et al., 2023; Jet Learn, 2024).
- *AI-Aided Tutoring:* Autism is not just a learning disability, it is characterized by behavioural, cognitive, and social difficulties in a child. Therefore, it becomes difficult for caregivers to single-handedly care for the child. With this, comes the role of AI, which not only assists in personalized learning but also creates a social robot for the child with whom they can interact and learn social skills and emotional intelligence without the fear of being judged (Bone & Smith, 2024).

While trying to comprehend the dynamic role of AI in teaching and learning, it is essential to shed light on the teaching pedagogies used for students with special education needs and/or disabilities (SENDs). One of the most commonly used pedagogic tools is the Targeted Individually Structured and Inclusive Intervention Programs for Students with Special Education Needs (TISIPfSENDs). The goal here is to divide the intervention into individual steps as per the appropriate level of readiness of the child, his/her interests, and provided autonomy; hence focusing on teamwork and learner centredness (Korea & Panagiotis, 2024). Based on such pedagogic tools for SENDs, educational institutions, research centres, and governmental initiatives have begun weaving AI into the existing curriculum (Song et al., 2024).

The next few paragraphs elaborate on such AI-driven platforms, tools, and services that have been proven to be effective in promoting inclusive education for students with SENDs:

- *Open AI's ChatGPT*: The most commonly known, accessed, and user-friendly AI platform is ChatGPT, which is based on machine learning and a large language model. Its multifaceted benefits have been reported in multiple studies. When it comes to assisting in diagnosing a concern, ChatGPT shows an encouraging success rate for neuropathological areas (Koga et al., 2023), is efficient for improving classification accuracy for autism spectrum disorders (ASDs) (Ullah et al., 2018; Liao et al., 2022), and is significantly accurate in identifying specific learning disorders (SLDs) like dyslexia, dyscalculia, and dysgraphia when used in parallel with existing diagnostic procedures (Ullah et al., 2018). The system also aids in understanding technical and complex subjects likes science, especially for students with ASD, SLD, and intellectual disability (Choi et al., 2023), improves productivity, motivation, verbal-expressive communication, student engagement and interest levels by promoting accessibility to learning materials, enhancing teamwork, foster personalized teaching, provides feedback, and acts as a virtual assistant teacher beyond the realm of the classroom (Alenezi et al., 2023; Jauhiainen & Guerra, 2023; Adiguzel et al., 2023; Baidoo-Anu & Leticia, 2023; Alshahrani, 2023; Rane, 2023). These benefits are not only applicable to students with SENDs, but also to the educators: the quality of personalized lesson plans is improved when made with AI chatbots and similar systems, they had more appropriate instructional and intervention goals, and teachers were better able to adapt pedagogical materials as per the needs of the learners and spent less time on creating assessments (Karakose & Tülübaş, 2023; Rakap, 2023). This easily accessible tool can aid in creating inclusive classrooms bridging the gaps in the teaching–learning process for varied learners. Researches have also highlighted the role of providing appropriate training to students and teachers, while at the same time incorporating AI into mainstream education (Song et al., 2024).

- *Intelligent Tutoring System (ITS)*: This is an AI-based software that focuses on the delivery of learner-oriented, adaptive tutoring systems. Its core structure includes working on a learner-centric model, which involves identifying the interests, strengths, and learning styles of the individual, delivering appropriate pedagogical content, and providing relevant feedback (Dutt et al., 2021). The software has been specifically successful in diagnosing and monitoring SLDs, cerebral palsy (CP), and developmental dysgraphia (DD) using its game interface, applications like TestGraphia, and an inbuilt kinematic assessment (Bortone et al., 2018; Mitrpanont et al., 2018; Tlili et al., 2020; Dimauro et al., 2020). Using human–computer interactions and augmented reality, ITS has been proven to enhance students' engagement, and improve learning outcomes, short-term memory and cognitive enhancement, and learner interactions with the educational content owing to its visually appealing nature (Ahuja et al., 2022). ITS builds confidence in learners with special needs as it aids in content absorption using mnemonics to improve their ability to process and retain information (Bobby, 2022). An important feature of the ITS is the inclusion of multidisciplinary teams that tailor solutions based on educational strategies (Gonzalez-Arias et al., 2021). It has also been tried, tested, and found to be of significant help in enhancing learning outcomes in children with hearing disabilities by fostering their social, personal, language and communication, time management, and technology-related skills (Megahed, 2019).

AI is changing the landscape of education in today's world, and with the fast-paced life and tough competition, students with special needs require AI's assistance to compete with the typical population. This calls for collaborative action between government agencies, school authorities, educators, parents, and these children to create, implement, and accept this revolutionary change. It is worthwhile noting that there are many case studies and examples of successful implementations of AI in educational settings and promoting inclusive environments (see Table 8.1).

There are apps, software, websites, portals, and other AI-based platforms prevalent in the education scenario. However, the majority of these projects are taken up in the Western world, suiting the West's economy, social, and legal structure. In order to extend India's current education system, and to make it more accessible to children with special needs, the government of India has laid down certain policies and programs. Some of the most important ones are discussed in the next section.

8.4 Brief Perspective on the Indian Government Initiatives

There are multiple acts, programs, and policies aimed at improving the education system in India which home in on diverse learners. The present section discusses a few of them in brief, while also adding certain recommendations of how AI can be assimilated in these initiatives.

National Education Policy (2020): This policy has specific provisions for children with special needs. Some of which include specialized teacher training for such learners, equipping them with appropriate skills and knowledge, designing the curriculum in a manner that it is accessible and relevant to all types of learners, creating and implementing enabling mechanisms such as language appropriate teaching materials, having assistive devices to ensure an inclusive classroom environment, and advocacy for the integration of children with special needs into mainstream classrooms to ensure participation in all academic and non-academic activities (Ministry of Human Resource Development, Government of India, n.d.). For this integration, the

TABLE 8.1

Case Studies and Examples of the Implementation of AI in Special Education

AI adopted	Institution/Author	Functioning	Impact
"Help Me See"	University of Alicante, Spain	Makes use of computer vision and machine learning to assist visually impaired students by identifying and narrating objects in the environment, allowing for a better navigation around the campus	Improved student engagement and participation in the campus activities (Digital Defynd, 2024)
"LinguaBot"	Beijing Language and Culture University, China	A natural language processing model, uses speech recognition technology to evaluate students' pronunciation and grammar in real time, offering them feedback to improve and personalize vocabulary exercises as per their pace	Improved language acquisition skills in non-native learners, as well as in students with difficulties in understanding grammar. It also improved class and community integration (Digital Defynd, 2024)
"Project Torino"	Initiative by Microsoft.	Helps visually challenged students to engage in coding through blocks, a tactile-based engagement, which is later digitally converted into screen-based codes	Helped visually impaired students to build problem-solving and computational thinking skills (SmartJobBoard, 2023)
"ATHYNOS"	Avila-Pesantez et al., 2018	An augmented reality game that focuses on improving mathematical skills for children with dyscalculia (a specific learning disorder); has components designed as per the capabilities of the player	Shown to improve mathematical reasoning in the children, specifically, basic numeracy, sequential ordering, and reasoning
"EMOTIFY GAME"	Rouhi et al., 2019	A digital application that makes use of audio features, such as pitch, tone, etc.; helps students with autism spectrum disorders to recognize feelings and express emotions	Increased engagement in the conversation; helps comprehend and re-learn effective behavioural and emotional responses towards others

TABLE 8.1 (Continued)

AI adopted	Institution/Author	Functioning	Impact
"KIWI Robot"	Arshad et al., 2020	An assistive technology that uses audio-video data (such as eye contact, verbal language), and supervised machine learning algorithms to adapt the lessons as per the needs of the learner (for autism)	It improved the social communication skills and mathematical skills of the learner

policy recommends the use of AI and other technologies, however, this may be of concern to some marginalized populations, who may not have access to the internet and other facilities. This may increase the existing inequalities, especially if the implementation of this policy is not supervised appropriately. The policy also seems to lack a fixed roadmap for the application of technologies that cater to special needs children, without further marginalizing them. Having said that, the policy suggests effective changes in the school system that are of benefit to learners, such as using MOOCs, online learning platforms like Swayam, etc., that allow learners to study at their own pace and time, and is bridging the gap between classroom and virtual learning.

Samagra Shiksha Abhiyan (2018–19): An overarching program initiated by the Ministry of Education, this aims at improving school education quality focusing on the two Ts: Teacher and Technology. It is comprised of three schemes: *Sarva Shiksha Abhiyan*, *Rashtriya Madhyamik Shiksha Abhiyan*, and Teacher Education. Some of its major interventions include inclusive education, digital initiatives, vocational education, strengthening of teacher education and training, to name a few (*Samagra Shiksha*, n.d.). This policy is an important step towards the provision of resources for children with special needs, such as the use of assistive devices like hearing aids, braille script, creating resource rooms, etc. However, the distribution of these resources needs to be monitored closely, while also ensuring effective training of the teachers to equip them to work with those devices and be sensitive towards the needs of special needs children.

Inclusive Education for the Disabled at Secondary Stage (2009–10): Under the scheme of *Rashtriya Madhyamik Shiksha Abhiyan*, this program provides assistance to special needs learners from IX standard onwards. It involves a student-centric component that caters to the medical and educational needs of the learners (such as therapeutic services, transport allowance, assistive devices, etc); and a teacher-related component that addresses the training requirements of teachers, the establishment of a barrier-free environment in the classroom for all learners, etc. (Nic, n.d.). The student-centric approach is a major highlight of this policy, which needs to be associated with mental health services being available to special needs children. This will aid them in not just improving their academic skills (as provided by the programs of the policy), but also to efficiently transition from special to mainstream classrooms.

National Policy for Persons with Disabilities (2006): This is a comprehensive framework for safeguarding the rights of and opportunities for people with disabilities. The focus here is on integrating these individuals into the mainstream education and employment scenario, but providing supportive measures such as legal aid (as backed by the Equal Opportunities, Protection of Rights and Full Participation Act of 1965), shifting from medical rehabilitation to social integration and rehabilitation, emphasis on and encouragement of self-help groups to promote entrepreneurship, reservation in government jobs under the People with Disabilities Act of 1995, and provision of skill-based vocational training for people with disabilities (Kumar & Government of India, 2006). Since the inception of this policy, researchers and educators have tried to analyse and critique it. One major issue with its implementation is the lack of awareness by the general population about the functioning of the policy and its benefits (Lekha, 2021). Secondly, the schools that are a part of this policy must make sure that learners with special needs are not directly placed into the mainstream classroom, rather, they must be first provided with additional support. This puts responsibilities on the shoulders of various stakeholders, and hence, their compliance must also be effectively monitored.

There are various other adjacent programs in addition to the aforementioned ones. It is worthwhile noting that almost all mention teacher training and building an inclusive classroom to make education accessible for all. Artificial intelligence can aid in developing specific programs for learners depending on their abilities, learning styles, interests, and previous progress. It can be used as a virtual tutor that supports the learning needs of the child even in the absence of an educator. Similarly, instead of teachers physically attending training programs, AI can be used to curate modules that can be accessed online, at one's own pace, and allow the educators to train with actual datasets/cases of children with special needs. AI can also bring in specialists and multidisciplinary professionals into the scenario, for instance, when dealing with an unusual case. This can aid in timely diagnosis, intervention, education, and finally follow-up for the cases. Special budgets or funding for the incorporation of AI-based assistance in a classroom can be a potentially benefiting step; especially, in a country like India with such a large demographic to cater to. Having AI-based assessment programs to evaluate whether the schemes are being implemented, and to have all relevant databases in a single (and multi-branched program) platform, which can be further used to expand the existing use of technology in education, would be of benefit.

8.5 Challenges and Ethical Considerations

In the revolutionary world of AI, it is essential to keep monitoring how one utilizes it and how challenging it can be to adapt to it, as it has become fundamental to contemporary existence (*5 Ethical Considerations for Using AI in Special Education*, n.d.). The role of AI has increased significantly, especially in breaking down the barriers regarding children with special needs. However, challenges have also arisen such as in terms of outputting data that may reflect the biases present in the repository data, which can lead to discriminatory outcomes for individuals with disabilities (Special Rapporteur on the Rights of Persons with Disabilities, 2023). Also, to promote its effectiveness in interpreting the data, there is a need for detailed analysis of preexisting datasets so that it can learn to expand its horizon (Rai et al., 2023). In a classroom setting, the educator also has an overseeing responsibility to be equipped to deal with varying abilities that presents challenges and also, they must distribute their attention between students with disabilities and those without, while ensuring that all students participate equitably and are treated fairly. Additionally, a lack of institutional resources and backing can hamper the implementation of inclusive educational practices and planning (Garg et al., 2020). This is of concern in educational settings where AI might treat students with disabilities as outliers, failing to accommodate their unique needs adequately. Its systems can inadvertently perpetuate biases present in historical data, leading to discriminatory outcomes against individuals with disabilities. The use of datasets in AI raises significant privacy concerns as well (De Micco et al., 2024). Ethical frameworks must be established to protect sensitive information, ensuring compliance with data protection regulations. These ethical concerns must be thoughtfully addressed to protect students and uphold the integrity of the educational systems. Safeguarding students' data and protecting the privacy and security of students' personal information is of utmost importance. Educational institutions also have a responsibility to educate students, parents, and staff about data privacy, the specific security measures being used, and their rights regarding their personal information. Transparency and clear communication are essential for building trust in AI systems and empowering students and families to make informed decisions about their data (Abramov, 2024).

It is also essential that a familiar and universal design be adopted for individuals with a diverse array of disabilities that incorporate their sensory, cognitive, and physical challenges to foster learning. Furthermore, the developer must ensure that the products are priced affordably (Shivani et al., 2024). On the technical side, AI systems require building comprehensiveness into the existing models and creating user interfaces that allow students and teachers to interrogate how the AI is arriving at its outputs (Disability Ethical AI, 2024), but beyond the technology itself, ongoing education and clear and accessible communication to students around AI systems is also essential. AI should be designed and deployed in ways that empower educators and enrich the learning experience but not as a wholesale replacement for human instruction and interaction (Sharma et al., 2024). Artificial intelligence enhances the efficacy of various methods and practices employed in teaching; nevertheless, if not implemented with care, we can foster a reliance on technology, instead of promoting the development of skills and strategies that encourage independence and self-advocacy among users in their educational pursuits (Shivani et al., 2024).

Some future directions and implications for policy makers, educators, and researchers, working in the realm of AI include the following:

- The implementation of the various programs requires regulatory channels to monitor the distribution of resources. At present, there are many government initiatives in place, however, the delivery of these programs may pose a significant challenge, keeping in mind the population and the varied demographic profile of India. People without internet access are likely to fall behind, and it is the responsibility of government authorities to address this issue and make arrangements to resolve this. Under certain schemes, there were provisions of laptops for learners, access to study material outside of the classroom, libraries, and resource centres, to name a few; however, it is a sad truth that many schools lack even the basic facilities of a classroom, leave alone having an inclusive classroom. These policies need major reforms and must be made to fit the economic backgrounds of all learners, while also ensuring the delivery of standard resources to all schools supporting learner-focused environments.
- In order to make available the AI-based resources, the government must make sure that there are effective training programs for teachers, students, and parents. It is necessary for educators to be aware of the functioning, benefits, and ethical concerns associated with the use of AI in the classroom settings, especially when it comes to using it for special education. This is to make sure that the learners are not further marginalized when it comes to using technological assistances. Secondly, there must be resource centres available for teachers as well, where they can perform hands-on training with the AI technologies, before implementing them in the classroom with the students. Parents as stakeholders also play an important role, and they must be involved in the process of implementing AI in the classroom. They can aid in the identification of the symptoms of their children, their triggers (especially with learners on the autism spectrum), the ways in which they perform best, etc., thereby helping schools to build a comprehensive, user-friendly system for learners. But for this to happen, parents need awareness and education programs too, while the school authorities must make sure to win their trust when it comes to data security in AI-based endeavours.

- There is a paucity in the Indian literature addressing the construction, implementation, and evaluation of AI-based technologies (apps/software/websites/student portals, etc.) for children with special needs. This calls for a multimodal approach, including qualitative researches (to understand the challenges and needs of learners), diagnostic studies (to analyse the effectiveness of AI in identifying problems, and making the technology sufficient to provide insights into the potential diagnosis of a disability), experimental researches (to enable the implementation of AI assistive technologies), longitudinal researches (to make sure that once the technology is in place, it yields productive results for learners); and so as to ensure a comprehensive understanding of the scope (and gaps) of implementing AI in the education setting.
- Collaboration between the IT sector and educationists is a crucial step towards achieving the technical excellence that is important for delivering AI-assisted programs for special needs children. Such future prospects should be kept as sustainable goals towards achieving inclusivity.

8.6 Conclusion

Incorporating AI in education is a powerful step ahead to mark the beginning of inclusivity and personalized learning for special needs children. Each student has varied needs and therefore has rights to access education; here, adaptive learning in AI plays a crucial role by providing education to all without compromising on quality. It allows educators to make the content more engaging, creative, and tailored for each student. A decade ago, AI seemed a distant future, an initiative that was far-fetched, but today we are living in that future, and this concept is no longer just to be found in books. Many universities from all over the world are developing software like Project Torino, ATHYNOS, EMOTIFY GAME, and KIWI Robot to help students with impaired vision, dyscalculia, and autism compete with the mainstream world. While AI is a powerful tool, it cannot replace human empathy and expertise, and this is where ethical considerations come into play. Students with special needs require compassion combined with education, therefore, maintaining a balance between judicial use of AI and human interaction is the only way forward to redefine education for students with special needs. Finally, for steady growth, AI must evolve in conjunction with government policies and the regulatory framework, ensuring a protective and trustworthy environment not just for students but also their parents, leading to a more sustainable synergy between the technology and human interaction.

References

Abramov, M. (2024, April 02). *Ethical Considerations in AI Model Development.* Keymakr. https://keymakr.com/blog/ethical-considerations-in-ai-model-development/

Adiguzel, T., Kaya, M. H., & Cansu, F. K. (2023). Revolutionizing education with AI: Exploring the transformative potential of ChatGPT. *Contemporary Educational Technology, 15*(3), ep429. https://doi.org/10.30935/cedtech/13152

Admin. (2024, September 4). AI in Special Education – Mantech Publications. *Mantech Publications Blog.* www.mantechpublications.com/blog/ai-in-special-education

Afzaal, M., Nouri, J., Zia, A., Papapetrou, P., Fors, U., Wu, Y., Li, X., & Weegar, R. (2021). Explainable AI for data-driven feedback and intelligent action recommendations to support students self-regulation. *Frontiers in Artificial Intelligence, 4,* 1-20. https://doi.org/10.3389/frai.2021.723447

Ahuja, N. J., Dutt, S., Choudhary, S. L., & Kumar, M. (2022). Intelligent tutoring system in education for disabled learners using human–computer interaction and augmented reality. *International Journal of Human-Computer Interaction,* 41(3), 1804-1816. https://doi.org/10.1080/10447318.2022.2124359

AI Grading Tool for Teachers | CoGrader | Use your own Rubric. (n.d.). AI website https://cograder.com/ai-grading

AI Tools for Teacher Grading — TOM DACCORD. (n.d.). TOM DACCORD. www.tomdaccord.com/ai-tools-for-teacher-grading

Alenezi, M. A. K., Mohamed, A. M., & Shaaban, T. S. (2023). Revolutionizing EFL special education: How ChatGPT is transforming the way teachers approach language learning. *Innoeduca International Journal of Technology and Educational Innovation, 9*(2), 5–23. https://doi.org/10.24310/innoeduca.2023.v9i2.16774

Alshahrani, A. (2023). The impact of ChatGPT on blended learning: Current trends and future research directions. *International Journal of Data and Network Science,* 7(4), 2029–2040. https://doi.org/10.5267/j.ijdns.2023.6.010

Arshad, N. I., Hashim, A. S., Ariffin, M. M., Aszemi, N. M., Low, H. M., & Norman, A. A. (2020). Robots as assistive technology tools to enhance cognitive abilities and foster valuable learning experiences among young children with Autism Spectrum Disorder. *IEEE Access,* 8, 116279–116291. https://doi.org/10.1109/ACCESS.2020.3001629

Atoui, H. (2024, July 29). *Enhancing Learning with AI-Powered Tutoring and Support Systems – London School of Science & Technology.* London School of Science & Technology. www.lsst.ac/blogs/enhancing-learning-with-ai-powered-tutoring-and-support-systems/

Attention-Deficit/Hyperactivity Disorder. (n.d.). National Institute of Mental Health (NIMH). www.nimh.nih.gov/health/topics/attention-deficit-hyperactivity-disorder-adhd

Autism Spectrum Disorder. (n.d.). National Institute of Mental Health (NIMH). www.nimh.nih.gov/health/topics/autism-spectrum-disorders-asd#:~:text=Autism%20spectrum%20disorder%20(ASD)%20is,first%202%20years%20of%20life

Baidoo-Anu, D., & Leticia, O. A. (2023). Education in the Era of Generative Artificial Intelligence (AI): Understanding the potential benefits of ChatGPT in promoting

teaching and learning. *SSRN Electronic Journal.* 7(1), 52-62. https://doi.org/10.2139/ssrn.4337484
Be10x, T. (2024, February 18). How AI tools are influencing education. *Be10 X.* https://be10x.in/blog/how-ai-tools-are-influencing-education/
Bobby, N. N. (2022). Mnemonic Technique Enhancing Learning Skills in School Students with Learning Disabilities. *International Journal of Research Publication and Reviews.* Volume number 3, issue number 5, page- 1032-1035
Bone, C., & Smith, C. (2024, February 1). Artificial Intelligence in Special Education | frog, part of Capgemini Invent. *Frog, Part of Capgemini Invent.* www.frog.co/designmind/artificial-intelligence-in-special-education
Bortone, I., Leonardis, D., Mastronicola, N., Crecchi, A., Bonfiglio, L., Procopio, C., Solazzi, M., & Frisoli, A. (2018). Wearable haptics and immersive virtual reality rehabilitation training in children with neuromotor impairments. *IEEE Transactions on Neural Systems and Rehabilitation Engineering, 26*(7), 1469–1478. https://doi.org/10.1109/tnsre.2018.2846814
Choi, E. (2023). A comprehensive inquiry into the use of ChatGPT examining general, educational, and disability-focused perspectives. *International Journal of Arts Humanities & Social Science, 4*(11), 1–7. https://doi.org/10.56734/ijahss.v4n11a1
Dass, N. M. M. (2024). A solution-based review paper on special needs/special education for schools in India. *GSC Advanced Research and Reviews, 19*(3), 182–190. https://doi.org/10.30574/gscarr.2024.19.3.0225
Datta, P. (2023). Special education needs: Disorder and disability in India in the light of National Education Policy 2020. *International Journal for Multidisciplinary Research, 5*(5). https://doi.org/10.36948/ijfmr.2023.v05i05.7157
De Micco, F., Tambone, V., Frati, P., Cingolani, M., & Scendoni, R. (2024). Disability 4.0: Bioethical considerations on the use of embodied artificial intelligence. *Frontiers in Medicine, 11, 1-9.* https://doi.org/10.3389/fmed.2024.1437280
DigitalDefynd, T. (2024, September 6). *Use of AI in Schools [25 Case Studies] [2025].* https://digitaldefynd.com/IQ/ai-in-schools-case-studies/
Dimauro, G., Bevilacqua, V., Colizzi, L., & Di Pierro, D. (2020). TestGraphia, a software system for the early diagnosis of dysgraphia. *IEEE Access, 8*, 19564–19575. https://doi.org/10.1109/access.2020.2968367
Disability Ethical AI. (2024, June 9). *Disability Ethical AI.* https://disabilityethicalai.org/
Dutt, S., Ahuja, N. J., & Kumar, M. (2021). An intelligent tutoring system architecture based on fuzzy neural network (FNN) for special education of learning disabled learners. *Education and Information Technologies, 27*(2), 2613–2633. https://doi.org/10.1007/s10639-021-10713-x
Garg, S., & Sharma, S. (2020). Impact of Artificial Intelligence in special need education to promote inclusive pedagogy. *International Journal of Information and Education Technology, 10*(7), 523–527. https://doi.org/10.18178/ijiet.2020.10.7.1418
Gonzalez-Arias, K., Galvez-Quezada, J., Robles-Bykbaev, V., Bermeo-Zambrano, F., & Leon-Pesantez, A. (2021). In Claudio da Rocha Brito, Melany M. Ciampi (eds). An intelligent decision support system for educational inclusion of children with disabilities: An approximation for multidisciplinary teams. In *2021 IEEE World Conference on Engineering Education (EDUNINE).* Institute of Electrical and Electronics Engineers Inc. (Vol. 12, pp. 1–6). https://doi.org/10.1109/edunine51952.2021.9429110

Haberstroh, S., & Schulte-Körne, G. (2019). The diagnosis and treatment of Dyscalculia. *Deutsches Ärzteblatt International*. 116, 107-114. https://doi.org/10.3238/arztebl.2019.0107

Jauhiainen, J. S., & Guerra, A. G. (2023). Generative AI and ChatGPT in school children's education: Evidence from a school lesson. *Sustainability, 15*(18), 14025. https://doi.org/10.3390/su151814025

Jet Learn. (2024, August 31). How AI Can Help Identify and Support Learning Disabilities. *Jet Learn*. www.jetlearn.com/blog/how-ai-can-help-learning-disabilities

Karakose, T., & Tülübaş, T. (2023). How can ChatGPT facilitate teaching and learning: Implications for Contemporary Education. *Educational Process International Journal, 12*(4), 7-16. https://doi.org/10.22521/edupij.2023.124.1

Koga, S., Martin, N. B., & Dickson, D. W. (2023). Evaluating the performance of large language models: ChatGPT and Google Bard in generating differential diagnoses in clinicopathological conferences of neurodegenerative disorders. *Brain Pathology*, 34(3), 1-4. https://doi.org/10.1111/bpa.13207

Korea, M. D., & Panagiotis, A. (2024). Factors arising from the utilization of Artificial Intelligence and large language models in special education and training. *European Journal of Special Education Research, 10*(2), 1-16. https://doi.org/10.46827/ejse.v10i2.5209

Kumar, M., & Government of India. (2006). National Policy for Persons with Disabilities. https://cdnbbsr.s3waas.gov.in/s3e58aea67b01fa747687f038dfde066f6/uploads/2023/10/20231013248779923.pdf

Lekha, K. S. (2021). Review of national policy for persons with disability, 2006. *ResearchGate*. www.researchgate.net/publication/350106479_REVIEW_OF_NATIONAL_POLICY_FOR_PERSONS_WITH_DISABILITY_2006

Liao, M., Duan, H., & Wang, G. (2022). Application of machine learning techniques to detect the children with Autism Spectrum Disorder. *Journal of Healthcare Engineering, 2022*, 1–10. https://doi.org/10.1155/2022/9340027

Megahed, F. (2019). Artificial intelligence applications and developing the life skills for students with special needs: A future look. *International Journal of Research in Educational Sciences, 3*(1), 175–194. https://doi.org/10.29009/ijres.3.1.3

Mehta, R. (2023). Education of children with special needs- A systematic review of literature. *Contemporaneity of Language and Literature in the Robotized Millennium, 5*(1), 1–5. https://doi.org/10.46632/cllrm/5/1/1

Ministry of Human Resource Development, Government of India. (n.d.). *National Education Policy 2020*. www.education.gov.in/sites/upload_files/mhrd/files/NEP_Final_English_0.pdf

Mitrpanont, J., Bousai, B., Soonthornchart, N., Tuanghirunvimon, K., & Mitrpanont, T. (Eds.). (2018). ICareADHD: A mobile application prototype for early child Attention Deficit Hyperactivity Disorder. In Jarernsri L. Mitrpanont,

Wudhichart Sawangphol (eds.). *2018 Seventh ICT International Student Project Conference (ICT-ISPC)*. Institute of Electrical and Electronics Engineers Inc. https://doi.org/10.1109/ictispc.2018.8523973

Nic, L. P. (n.d.). *Inclusive education of the disabled at secondary stage | Government of India, Ministry of Education*. www.education.gov.in/iedss

Pardamean, B., Suparyanto, T., Cenggoro, T. W., Sudigyo, D., & Anugrahana, A. (2022). AI-based learning style prediction in online learning for primary education. *IEEE Access, 10*, 35725–35735. https://doi.org/10.1109/access.2022.3160177

Rai, H. L., Saluja, N., & Pimplapure, A. (2023). AI and learning disabilities: Ethical and social considerations in educational technology. *Educational Administration: Theory and Practice*, 29(4), 1–8. https://doi.org/10.53555/kuey.v30i5.4263

Rajagopal, A., Nirmala, V., Jebadurai, I. J., Vedamanickam, A. M., & Kumar, P. U. (2023). Design of generative multimodal AI agents to enable persons with learning disability. In Elisabeth André, Mohamed Chetouani, Dominique Vaufreydaz, Gale Lucas, Tanja Schultz, Louis-Philippe Morency, Alessandro Vinciarelli (eds.). *ICMI '23 Companion: Companion Publication of the 25th International Conference on Multimodal Interaction*. Association for Computing Machinery, New York, (Vol. 40, pp. 259–271). https://doi.org/10.1145/3610661.3617514

Rakap, S. (2023). Chatting with GPT: Enhancing individualized education program goal development for Novice special education teachers. *Journal of Special Education Technology*. 39(3). https://doi.org/10.1177/01626434231211295

Rane, N. (2023). Chatbot-enhanced teaching and learning: Implementation strategies, challenges, and the role of ChatGPT in Education. *SSRN Electronic Journal*. https://doi.org/10.2139/ssrn.4603204

Roessingh, J. J., Poppinga, G., Van Oijen, J., & Toubman, A. (2019). Application of artificial intelligence to adaptive instruction – Combining the concepts. In Robert A. Sottilare; Jessica Schwarz (eds.). *Lecture Notes in Computer Science*. Springer, Cham, (pp. 542–556). https://doi.org/10.1007/978-3-030-22341-0_43

Rouhi A., Spitale M., Catania F., Cosentino G., Gelsomini M., & Garzotto F. Emotify: Emotional game for children with autism spectrum disorder based-on machine learning. In Geof Mulgan, Scott Page (eds.). *Proceedings of the 24th International Conference on Intelligent User Interfaces: Companion; Marina del Ray, CA, USA. 16–20 March 2019*. Association for Computing Machinery, (pp. 31–32).

Samagra shiksha. (n.d.). https://samagra.education.gov.in/about.html

Sharma, S., Singh, G., Sharma, C. S., & Kapoor, S. (2024). Artificial intelligence in Indian higher education institutions: A quantitative study on adoption and perceptions. *International Journal of Systems Assurance Engineering and Management*. https://doi.org/10.1007/s13198-023-02193-8

Shivani, N., Gupta, M., & Gupta, S. B. (2024). A systematic analysis of AI-empowered educational tools developed in India for disabled people. *Information Technologies and Learning Tools*, *100*(2), 199–216. https://doi.org/10.33407/itlt.v100i2.5501

SmartJobBoard. (2023, December 7). *Case Studies of AI-Empowered Inclusive Education Initiatives*. DiverseJobsMatter. https://diversejobsmatter.co.uk/blog/case-studies-of-ai-empowered-inclusive-education-initiatives/

Snowling, M. J. (2005). Specific learning difficulties. *Psychiatry*, 4(9), 110–113. https://doi.org/10.1383/psyt.2005.4.9.110

Song, Y., Weisberg, L. R., Zhang, S., Tian, X., Boyer, K. E., & Israel, M. (2024). A framework for inclusive AI learning design for diverse learners. *Computers and Education Artificial Intelligence*, *6*, 100212. https://doi.org/10.1016/j.caeai.2024.100212

Special Rapporteur on the Rights of Persons with Disabilities. (2023, May 29). *Artificial Intelligence and the Rights of Persons with Disabilities*. Heba Hagrass. https://srdisability.org/thematic-reports/artificial-intelligence-and-the-rights-of-persons-with-disabilities/

Stefanic, D. (2024, June 12). *AI and Learning Styles: Tailored Education Solutions*. Hyperspacemv – the Metaverse for Business Platform. https://hyperspace.mv/learning-styles-ai/

Store, H. O. (2024, March 8). The Future of AI in Education: Personalized Learning and Beyond. *Miscellaneous – Tech Takes*. www.hp.com/in-en/shop/tech-takes/post/ai-in-education

Tlili, A., Najjar, R., Essalmi, F., Jemni, M., Chang, M., Huang, R., & Chang, T. (2020). Unobtrusive monitoring of learners' game interactions to identify their dyslexia level. In Maiga Chang, Demetrios G. Sampson,

Ronghuai Huang, Danial Hooshyar, Nian-Shing Chen, Kinshuk, Margus Pedaste (eds.). *2020 IEEE 20th International Conference on Advanced Learning Technologies (ICALT)*. Institute of Electrical and Electronics Engineers. https://doi.org/10.1109/icalt49669.2020.00040

Ullah, K. R., Lee, J. A. C., & Oon, Y. B. (2018). Machine learning and dyslexia-Diagnostic and Classification System (DCS) for kids with learning disabilities. *International Journal of Engineering & Technology*, *7*(3.18), 97–100. https://ir.unimas.my/id/eprint/21523/

UNESCO. (2023, February 6). *Ethics of AI: Challenges and Governance* [Video]. YouTube. www.youtube.com/watch?v=VqFqWIqOB1g

Index

FSC
www.fsc.org
MIX
Papier | Fördert
gute Waldnutzung
FSC® C083411

Kolibri 360 GmbH
Ferdinand-Jühlke-Straße 7
99095 Erfurt, Deutschland
produktsicherheit@kolibri360.de